Astronomers' Universe

Other titles in this series

Origins: How the Planets, Stars, Galaxies, and the Universe Began (forthcoming)
Steve Eales

The Future of the Universe (forthcoming)
A.J. Meadows

Frank Levin

Calibrating the Cosmos

How Cosmology Explains Our Big Bang Universe

 Springer

Frank S. Levin, PhD
Professor Emeritus of Physics
Brown University
Providence, RI 02912
USA

Jacket illustration: WMAP image of the anisotropies in the cosmic microwave background radiation, shown on an oval projection that represents the full sky. Courtesy of NASA and the WMAP science team.

Library of Congress Control Number: 2005937514

ISBN-10: 0-387-30778-8
ISBN-13: 978-0-387-30778-7

Printed on acid-free paper.

9 8 7 6 5 4 3 2 1

springer.com

Preface

Calibrating the Cosmos is based on lectures I gave in several adult-education courses. By dealing with technical details in a descriptive way, I structured the courses for people who wanted to gain some knowledge of the physical universe as it is currently understood but had neither a science nor a mathematics background. The lectures were hard science, softly presented. As with the course, so with the book: it is written for persons whose curiosity about the physical universe extends to a readiness to learn some of the relevant technical aspects, presented descriptively.

Although many primary astronomical and cosmological experiments are identified, my emphasis is on the assumptions and theoretical concepts that underlie the measurements and the efforts to interpret and understand the resulting data. Taken together, observational information and theoretical ideas about the Universe form an elegant intellectual tapestry. I have treated some of its threads only cursorily; for instance, the history of astronomy, white dwarf stars, black holes, and the theory of inflation. Some experiments on the cosmic microwave background radiation are omitted entirely, though they are indirectly referred to. Not all the sources are given for the numbers I quote. In addition, for reasons stated in Chapter 6, the question of structure (e.g., the distribution of galaxies or clusters of galaxies) is omitted entirely.

The "details" of technical details are also items that I have glossed over, usually in favor of broadly constructed descriptions. Numbers, however, are not technical details! They are indispensable elements in describing and characterizing a Universe that is known to be very BIG now but is believed to have started out very small at very tiny times. Numbers of various kinds are sprinkled throughout the book, some in the text, some in tabular form.

To help make some observational/theoretical information easier to grasp, I have followed standard practice and portrayed it graphically. In doing so, I have assumed that graphical-type representations are no more difficult to understand than the graphs or curves that describe the behavior of the usual stock-market indicators. Furthermore, many descriptions are illustrated by simple line drawings, each designed to enhance your comprehension of a particular written analysis.

Although the book is intended for readers without a science or math background, it does contain a few equations, including Einstein's famous formula $E = Mc^2$. I have tried to explain in plain English the meaning of each of the equations and proportionalities.

Theoretical cosmology is able to generate many different "universes." To distinguish them from "our" universe, I refer to ours as the *Universe*, while those generated by theory I denote *universe* with a lowercase first letter. In other words, *a* universe as opposed to *the* Universe. In analogy to using *Universe* and *universe*, I similarly use *galaxy* to refer to a very large aggregation of stars held or bound together gravitationally, whereas *Galaxy* always signifies our own, the Milky Way Galaxy.

By the way, a long-standing question has been which, if any, of the many theoretical universes most closely corresponds to the Universe. It is likely that the answer to this question is close at hand; one of the pleasures for me in lecturing on and writing about our Universe is describing the "how" of the answer.

I am aware that emphasizing theory and theoretical concepts could pose a risk: unlike the course attendees, you cannot query me should you fail to grasp an idea or an explanation. To help minimize this risk, persons with widely varying backgrounds—the majority of whom were not technically trained—agreed to read and comment on portions of the book in draft form. Their valuable critiques have led to improvements in both the writing and the content. Naturally, any errors or infelicities that remain are my sole responsibility.

Finally, let me draw your attention to two other aspects of the book. First, it is divided into the same two portions that formed the syllabus for the adult-education courses. The first part, ending with Chapter 4, contains the background material that I thought would help the adult students in my courses understand the second portion, which concentrates on cosmology and the Universe. I hope readers of this book will find this arrangement to be beneficial as well.

The other aspect is the set of Web sites listed in the Bibliography. They not only provide ancillary material but also can function as information sources for those of you who would like to remain up to date. You might even have fun logging onto your favorite Web browser and exploring the sites produced by searching phrases such as *Big Bang, black holes, dark matter, gravitational lensing, supernovas, CMB, WMAP, dark energy, inflation,* and so forth. If reading this book encourages such activity, one of its goals will have been met. Happy reading and browsing!

[Note added in proof: In March 2006, the Wilkinson Microwave Anisotropy Probe (WMAP) team announced their first results since 2003. Included is a revised estimate of when stars were first formed (see Figure 25), and confirmation of another prediction of inflation theory, thereby adding further support for this very early Universe scenario (described in Chapters 7 and 9). Some details of the new results, which also contain an updated map of the hot and cold spots of the Universe and an extrapolation to a trillionth of a second after the Big Bang, can be found at the WMAP Web site, listed in the Bibliography.]

Contents

1. Introduction: The Splendid Science

Cosmology! The branch of knowledge concerned with the origin, evolution, and properties of the Universe, cosmology is arguably the grandest of human endeavors, for what could be grander than attempting to understand the cosmos? The quest to achieve this understanding is ancient. Its unknown origin dates back thousands of years, when people in different cultures recorded the regular motions of the planets and stars and then used their observations to create calendars, to predict celestial events, and to speculate on the origin of the cosmos.

Although the science of astronomy got its start from these venerable beginnings, cosmology itself has emerged only recently as a branch of science in the modern sense of the word. Its emergence was due in large part to the accidental discovery in 1964 of a type of radiation known as the *cosmic microwave background radiation*, now referred to by the acronym *CMB*. Once the significance of the *CMB* was understood and publicized (and I'll explain it shortly), more and more people started to do research on cosmological topics, a community was formed, textbooks were written, and the new discipline of cosmology gradually came into being. It has become one of the most bountiful of the sciences.

Cosmology's stunning revelations fall into one of two categories: theoretical or observational/experimental. Among the most important theoretical investigations is the study of model universes, especially the ones produced by the universe-generating, mathematical theory known as general relativity. Model-universe studies began soon after Albert Einstein's paper on general relativity appeared in 1916. Prior to the 1960s, however, and despite similarities between some of them and our own Universe, model-universe studies generated relatively little interest among most scientists. This was due in part to the excitement created by new research areas such as nuclear physics. Equally important, if not more so, was the mistaken perception that experiments could not connect any of the model universes with our own Universe.

This perception was dramatically altered by the serendipitous discovery of the CMB. Curiously enough, the discovery was made by two radio astronomers working for the Bell Telephone company! (See Chapter 6 for more details.) The microwave background radiation was quickly understood to be a previously predicted type of radiation that characterizes the early history of an entire class of theoretical universes, thereby providing the previously missing connection.

The existence of the CMB implies that our Universe is a member of the class of theoretical universes described by *Big Bang cosmology*. "Big Bang" refers to a generic type of expanding universe that has evolved from an explosive event, although the phrase itself was initially meant to be disparaging. It was introduced by a proponent of a theory known as *steady state cosmology*. Rather than evolving from an explosive event, the theoretical universes of steady state cosmology exist essentially unchanged in time, having neither a beginning nor an end. However, the CMB can only be accommodated in the steady state scenario by means of *ad hoc* assumptions, whereas it is a natural ingredient of the Big Bang framework.[a] Big Bang cosmology has triumphed, becoming a new paradigm, and the phrase *Big Bang* is now well-known outside of scientific circles. The discoverers of the CMB were awarded the Nobel Prize for a discovery that proved to be one of the most consequential of the 20th century.

Suppose that the CMB had not been detected, but that the Universe was somehow known to be a member of the Big Bang class of universes. This would necessitate its containing the CMB—which it does. But because the Universe is clumpy—apart from radiation, it is mostly empty space sparsely populated by galaxies and various other objects—theory predicts that the background radiation must also be clumpy. That is, the CMB measured from one region of the sky should differ slightly from the CMB when

---

[a]An *ad hoc* assumption or theory explains one fact only. It is scientifically unsatisfactory because it has no predictive power and therefore cannot be tested.

measured from any other part of the sky. If these differenccs were to exist, they would mean that the CMB deviates from perfect uniformity. Were such a deviation found, it would be dramatic evidence for the existence in the early Universe of the tiny nonuniformities in the distribution of matter that eventually led to galaxy formation.

The predicted nonuniformity, known as the *anisotropy* in the CMB, aroused great interest in the cosmology/astronomy communities. It led to the launching, in the late 1980s, of a satellite bearing equipment designed to detect the anisotropy. Called the cosmic background explorer and abbreviated COBE, it obtained data in 1992 that verified the prediction. The measured anisotropy was about 1 part in 100,000, or a thousandth of a percent, very small but much larger than experimental uncertainty.

The miniscule size of the anisotropy is a feature of the utmost significance, for hidden in it are clues that, suitably interpreted, yield information about the large-scale behavior of the Universe. Such information includes the overall geometry of the Universe, the amounts of both the luminous and the nonluminous, or "dark," matter in it, and the strength of the quantity (discussed in Chapter 6) that Albert Einstein once referred to as his "greatest blunder." The anisotropy's hidden treasures have motivated a host of theoretical investigations and experimental measurements. Highly accurate data have been obtained from many experiments carried out after the COBE mission. Notable among these investigations are those carried out by the Wilkinson Microwave Anisotropy Probe (WMAP) and the Sloan Digital Sky Survey (SDSS), discussed in Chapter 7.

The WMAP and SDSS findings, first reported in 2003, have been the best sources for evaluating the quantities that I denote the *parameters of the Universe*. Defined within the context of Big Bang cosmology, these parameters uniquely specify many properties of our Universe.

That these parameters, which are derived from theory, actually *can* specify properties of our Universe is based on the widely held belief of cosmologists that our Universe is uniquely identified with a theoretical universe generated by Big Bang cosmology. Underlying this identification are the facts that both our Universe and members of a particular class of Big Bang universes are each expanding, contain the CMB, and are *homogeneous* and *isotropic.*[b] The sharing of these common features is evidence that not only is there a unique relation between our Universe and a member of the class of Big Bang universes but also that they behave in the same way. Knowledge of one thus provides information on the other.

To learn which of the theoretical universes correlates with ours requires deducing the parameter values from measurements made, for example, on supernovas, on the CMB and on galaxies, and then inserting these values into the relevant theoretical formulas. Such an insertion will select the theoretical universe to which ours corresponds, while from it, properties and the behavior over time of our Universe can be determined.

A key aspect of the theoretical analysis, indeed, one of the most astonishing in all of modern cosmology, is that the past, present, and future size of our three-dimensional Universe is obtained from just one quantity! This single quantity is known as the *universal scale factor*, and its existence is a consequence of the homogeneity and isotropy properties that the Universe enjoys in the large. In Chapter 6, I'll explain why the scale factor exists, and in Chapter 7 I'll discuss the time evolution of the model universes generated by the scale factor.

While the size of the Universe over time is described by the scale factor, the scale factor depends on the values of the parameters. Thus there is an exquisite linkage between the CMB and the time behavior of the Universe. The parameters and scale factor play crucial roles in elucidating other aspects of the Universe, discussed in Chapters 7 and 8.

The scale factor is related to one of the most important quantities in cosmology, the Hubble constant, first identified and estimated by the astronomer Edwin Hubble. He showed that our Universe is expanding in such a way that the speeds with which galaxies are receding from the earth are proportional to their distances away from it; the proportionality constant in this relation is the one that bears his name. Although the Hubble constant itself

---

[b]I'll define and illustrate the terms *homogeneous* and *isotropic* in Chapter 6, but you can look them up now in the Glossary, which defines the other technical terms I use in this book.

is highly significant, the relation it enters is equally so, as I will show later. Known as Hubble's law, this latter relation is also a consequence of homogeneity and isotropy.

Without knowledge of the distances to galaxies beyond the Milky Way as well as their recession speeds, Hubble could not have deduced his law. Accurate measurement of astronomical distances, and later of cosmological ones, has been an essential requirement in all scientific attempts to understand the Universe, and you may well have wondered how such measurements have been accomplished. The answer is through a set of interlocking methods that form a hierarchy, one in which the easier-to-obtain shorter distances become the springboard for reaching out to longer distances. This collection of methods is known as the *cosmic distance ladder*, and although only the lower rungs were available to Hubble, they sufficed—spectacularly well—for his purposes.

Distance determination is so vital to astronomy and cosmology that *parallax*, the lowest method on the distance ladder, is the main subject of Chapter 2. When parallax fails, some of the methods that supercede it rely on the properties of certain types of stellar phenomena, for example *Cepheid variables* and *type Ia supernovas*. The role played by these exotic entities in determining distances is one of several reasons for my including the very broad discussion of stars of Chapter 4; another is the intrinsic interest that stars hold for most persons, especially stellar end-stages such as white dwarfs and black holes. Furthermore, stars shine: they are the most populous of the luminous ingredients in the Universe, and gaining some understanding of them is an essential element in appreciating the cosmos.

Hubble not only needed reliable distances, he had to know the recession speeds of the galaxies. They were—and are—obtained using a mechanism that exploits the wave properties of light and radiation. An essential element in understanding the cosmos is grasping how scientists have deduced that galaxies are receding from one another, as well as how fast they are moving away. This alone is a powerful reason for my reviewing light and radiation in Chapter 3. Another is the fact that light, and radiation in general, is the sole source of observational information about the Universe (hearing, taste, and smell obviously don't work!).

The cosmic microwave background is an almost perfect example of a kind of radiation known as *blackbody*. It is crucial to the construction of a timeline for the Universe that the CMB is of this type, as I discuss in Chapter 8. And, not only is the CMB approximately blackbody in nature but so also is the radiation emitted by the sun. The interplay between these very different entities neatly illustrates both the unity of the Universe and the use of terrestrial science to explain it.

Of course, not everything one wishes to know about the cosmos has been or can be deduced by examining its various forms of radiation. Critical aspects of it remain unknown, for instance, the identity of the nonradiating dark matter as well as the nature of the *dark energy* causing the expansion of the Universe to accelerate. (This acceleration is another 20th-century discovery that has revolutionized thinking about the cosmos.) Moreover, cosmology is not yet a completely fleshed-out science, so that explanations of some observational or inferred phenomena are based on conjectures that range from the highly likely to the highly speculative (see Chapter 9).

Even though not all the answers are in, much has been ascertained. Thus, while the nature of dark matter remains a mystery, the relative amounts of the current contents of the Universe and the nature and times of occurrence of many events that took place during its evolution have been estimated. Its large-scale geometry is known. An analysis of WMAP data combined with those from other experiments leads to the time of the Big Bang as approximately 13.7 billion years ago. The diameter of the visible Universe can also be estimated. It is roughly a quarter of a million billion billion kilometers $(0.25 \times 10^{24}\,\text{km})$,[c] or a sixth of a million billion billion miles. As you will discover in this book, these and other results, along with some of the conjectures about the cosmos, are as astonishing as any that occur in a non-cosmological context: the Universe *is* comprehensible, and cosmology explains much of it.

---

[c]The power-of-ten notation, for example, 10^{24}, is described in Appendix A.

2. Measuring Distances: On the Earth, in the Solar System, to the Nearby Stars

Distances play much the same role in astronomy and cosmology as perspective does in landscape painting: change either of them and the resulting picture changes. Accurate distances are required if you are to obtain a reliable picture of the Universe, just as they are in determining the size of the earth or the solar system or the Galaxy. The problem in each of these instances is the same: how is the requisite distance to be obtained when a direct measurement cannot be made? The solution is through the use of indirect methods, and I begin the description of them in this Chapter.

Attempts to measure distances, both successful and not, are part of the history of astronomy. Many of the successful procedures have been organized into a hierarchy known as the *cosmic distance ladder*, with each rung describing a distinct method, the lower ones typically supporting the higher ones. As one climbs the ladder, the associated distances increase; unfortunately, all of the procedures are imprecise, so that the inaccuracies of the lower-rung methods are incorporated into those of the higher rungs. Because inaccuracy is an inevitable aspect of this enterprise, great efforts have been made to ensure high precision in the shorter-distance measurements. I shall consider aspects of errors after introducing appropriate distance units.

The main distance method examined in this Chapter is denoted *parallax*. It occupies the lowest rung on the ladder and extends only to the "nearby" stars. Although such distances are small on the cosmological scale, there are two reasons for beginning with parallax: first, its assumptions and its *one-angle/one-known-distance* characteristic can be exposed in the more familiar setting of certain types of terrestrial measurements; second, it has been applied in the solar system. The former is important because the assumptions, which are rarely identified, are not all valid in the case of cosmological measurements. The latter application is useful because I will use it as the platform for discussing some concepts and details of the solar system such as planetary orbits, as well as mass and density, quantities essential for describing the Universe both observationally and theoretically.

Parallax (also known as *trigonometric* parallax) employs a *measured angle* and a *predetermined length* to evaluate the desired stellar distance. These same two elements entered the first measurement of the earth's radius, carried out ca. 240 BCE by the Greek philosopher Eratosthenes, a one-time director of the renowned library in Alexandria, Egypt. However, the one-angle, one-known-length method is not limited to measurements of very large lengths: it can also be used to determine quite ordinary distances, such as the heights of fixed vertical objects like telephone poles, trees, or sailboat masts. Since it is simplest to explain the method for this latter class of objects, I'll introduce the discussion of parallax by describing a procedure for measuring the height of a standing telephone pole without climbing it. A key element will be identifying the relevant assumptions. After that, I'll go on to the method used by Eratosthenes.

Measuring the Height of a Standing Telephone Pole

Figure 1 is a schematic depiction of a telephone pole, whose height h is to be determined. To begin, one marks off a length D to the left of the pole; it is the predetermined-distance portion of the method. The angle-measuring device, e.g., a protractor or similar instrument, shown as the small circle with a plus sign (+) in it, is then put into the ground at this distance. By creating a line of sight from the center of the protractor to the top of the pole, indicated in the figure by the dotted line, the measurer defines an angle, labeled A, whose value (in degrees) is read off the protractor. The telephone pole, the distance D, and the dotted line form a triangle, which is a shape from plane, or flat-space, geometry—the geometry of Euclid.

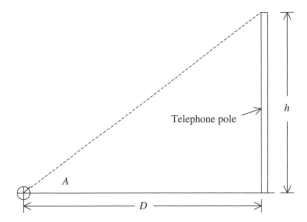

Figure 1. Illustration of the one-angle, one-measured-length method of determining a distance. To "measure" the height h of the telephone pole, one needs only to measure the preselected distance D and the angle A, where the symbol $\oplus$ represents an angle-measuring device such as a protractor.

h is uniquely determined by this construction. Of course, only D and A are measured: h itself is not. Instead, its numerical value is obtained from the other two measurements by using a mathematical formula based on plane geometry.[1] Nevertheless, the method is referred to as the *measurement of h*, just as the method of parallax is a means of "measuring" a stellar distance. Each is an example of the indirect procedure I mentioned at the beginning of the Chapter.

Although the foregoing description may seem straightforward, it contains some unspecified but crucial assumptions. First, by creating and using a triangle to define A, the geometry of flat surfaces (Euclidean geometry) is assumed to be valid. And indeed it is, as long as the distance D is not so great that the curvature of the earth's surface needs to be taken into account. This is normally the case because in any small region—that is, locally—the curvature is far too small to change the geometry from planar to spherical. But if the curvature were to become noticeable, then spherical geometry might become necessary. Spherical geometry would call for a different math formula, since the relevant distance would not be a straight line but a portion of a great circle, the type of route followed by airplanes flying long distances or ships crossing oceans.

A second assumption is that the pole and the protractor remain stationary, so that neither h nor D changes. For the type of measurement described above, this may seem like a frivolous remark, but in an expanding universe, quantities analogous to D *are changing*, and one must take care in dealing with distance. A third unstated assumption is that both the distance D and the angle A not only can be measured, but that it can be done with an accuracy sufficient for the purpose at hand. As I note later in this Chapter, for D large enough, the uncertainty in angles can become significant, whereas for most astronomic and cosmological distances it is impossible even to discern a parallax angle. When this occurs, parallax must be replaced by another method, one from a higher rung of the cosmic distance ladder.

An Aside on Angles

Since the procedure just outlined involves angles, let us take a small detour away from the next measurement—that of the earth's radius—and focus attention on the units in which angles are specified. In nontechnical applications, angles are measured in *degrees* and are indicated by the symbol ° placed as a superscript to the right of the numerical value; for instance, 30°. The degree is a concept originally associated with circles and was formulated by mathematicians of the ancient Babylonian civilization. Rather than 10, the base of the decimal system, they favored the number 60 and its multiples and divisors. In particular, they divided the circumference of a circle into 360 equal segments of arc and then defined the angle between the two radii drawn to the ends of one such segment as equal to 1°. In other words, one such arc segment *subtends* an angle of 1°. This division of a circle's circumference into 360 segments means that there are 360° in a circle. An arbitrary angle defined in this way (not equal to 1°) is shown in Figure 2.

An angle may not always be expressible as an integer number of degrees: its value may involve a fraction of a degree. The Babylonians dealt with this possibility by dividing each whole degree into 60 equal parts called *minutes*, indicated by the prime symbol, so that 1° contains 60'. And just as a degree comprises 60 minutes,

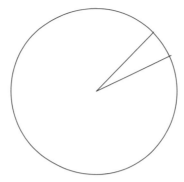

Figure 2. An angle subtended by a short arc of a circle and contained between two radii.

a minute was divided into 60 smaller portions, denoted *seconds*. The symbol for a second is the double prime, for instance 30″, which is one half of a minute. The Babylonians ended their subdivisions here; for smaller subdivisions, the modern decimal system is used, for example one tenth of a second is written 0.1″. Such small values are common in parallax measurements. Of course, one can avoid minutes altogether, replacing them with tenths of a degree (see below).

These Babylonian subdivisions into sets of 60 define time units as well: 60 minutes in an hour and 60 seconds in a minute. Despite their lacking the advantages of a decimally based set of units—the division of the day into 24 one-hour portions is an ancient Egyptian construction—the Babylonian/Egyptian system remains in effect today and is highly unlikely to be replaced: usage and tradition have trumped numerical convenience.

Back to Yesteryear: Measuring the Earth's Radius

When Eratosthenes measured the earth's radius, the cosmology he believed in was that of Aristarchus (ca. 320–250 BCE), which held (almost correctly!) that the earth was spherical,[2] that it rotated on its own axis, and that it revolved about the sun. Eratosthenes also believed, as we do not, that the earth was embedded in a spherical shell that did the actual revolving. That the earth is revolving about the sun and not *vice versa* is the hallmark of a heliocentric cosmology.[a]

Eratosthenes had learned that at noon on the summer solstice in Syene (the Egyptian city now called Aswan), sunlight cast no shadow (it fell perpendicular to the earth's surface there). However, at noon on the same day in Alexandria, sunlight fell obliquely on the earth's surface, making an angle of about 7.2° with an upright stick. Using these facts and the then known distance of 5000 *stadia* between the two cities, he was able to calculate the value of the earth's radius (he "measured" it), expressing the result in stadia. (To convert his result to miles or kilometers, and thus determine its accuracy, one needs to know how many stadia there are in a mile or a kilometer, a point considered shortly.)

Recalling comments I made above, it might seem that a spherical earth would have required Eratosthenes to base his calculation on spherical geometry. It was not needed because the center of the earth and the two cities lie in a plane: Euclidean geometry sufficed. The formula he used relates an arc of a circle to the angle it subtends at the center and to the radius. In the case at hand, the arc length is the distance between the two cities, and the radius is that of the earth. The former is the known length of the method; the angle he needed is the one subtended at the center of the earth by the ends of the arc. Because he couldn't measure it directly, Eratosthenes replaced it by an angle of equal size that he *could* measure.

The procedure that led to the desired angle is based on Figure 3, which shows a cutaway portion of the earth, defined by the plane noted above. It passes through the earth's center and the two cities on the surface; the drawing is not to scale. Depicted in the figure are arrows denoting the parallel set of the sun's rays, which strike the earth's surface perpendicularly at Syene and obliquely at Alexandria, plus the radius from the earth's center to each city

---

[a]Opposing it was the geocentric cosmology of Aristotle and others, codified in the second century by Claudius Ptolemy through publication of his book *The Almagest*. Geocentricity held sway in Europe for well over a thousand years, until Nikolaus Copernicus challenged it in the late 1500s.[3]

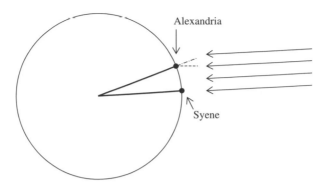

Figure 3. Illustration of the geometry used by Eratosthenes to "measure" the earth's radius. The circle denotes a plane cut through the center of the earth; the left-pointing arrows represent the parallel rays of the sun striking the earth's surface; the heavy lines are the radii to the cities of Alexandria and Syene, each symbolized by a heavy black dot; the dot-dash line is an extension of the radius to Alexandria; the dashed line is parallel to the radius to Syene.

(in bold), with the one to Syene parallel to the sun's rays. The dashed line drawn at Alexandria is parallel to both the sun's rays and the radius to Syene, and the vertical stick is represented by the dash-dot line, which is an extension of the radius to Alexandria.

There are two angles in the figure: the subtended one between the two radii—which Eratosthenes needed to know—and the one between the dot-dash and dashed lines. The key relation for him was the plane-geometry result that these two angles are equal. But, the angle between the dot-dash and dashed lines is the *same* as the angle with which the sun's rays strike the ground at Alexandria, viz., 7.2°. Hence, the subtended angle between the two radii is also 7.2°.

This conclusion, plus the fact that arc length is the product of the radius and the subtended angle (7.2°), enabled Eratosthenes to calculate the radius of the earth. He found it to be approximately 39,788 stadia, corresponding to a circumference of the earth of 250,000 stadia. His measurement of one angle and one distance plus his use of plane geometry yielded a "measurement" of the earth's circumference and radius.

The method just outlined is a triumph of the intellect, an imaginative use of reasoning based on analogy and geometry.

Although it is incisive, one must ask if the result is accurate. The only way to answer such a question is to convert stadia to contemporary units and then compare with the modern value. Assuming that the contemporary distance of 500 miles between Alexandria and Aswan is the same as the ancient Alexandria–Syene distance of 5000 stadia (however that value was obtained in the third century BCE), then one *stade* (the singular of stadia) equals one tenth of a mile. Hence, Eratosthenes's values convert to 25,000 miles for the earth's circumference and about 3979 miles for its radius.

How do these numbers compare with the contemporary values? Unfortunately, the comparison is not straightforward because the earth's circumference and radius are not uniquely defined! On the one hand, and even forgetting the existence of mountains, the earth is not spherical: its surface contains a variety of small deformations superimposed on one another, including a slight pear shape. On the other hand, even if this latter fact were to be ignored—as it can be for the present purposes since these deformations are small—there is the problem of the earth's bulge: due to its rotation, the earth is fatter at the equator than elsewhere (cf. note 2). The standard solution to this non-unique-radius problem is to use the equatorial value, in which case the earth's "radius" is found to be 3963 miles or 6378 km, and its circumference is almost identical to 25,000 miles or 40,074 km. The agreement between these values and the measurement of Eratosthenes is excellent. Not only is his result remarkably accurate, it was accepted as correct by his contemporaries, thereby demonstrating the esteem in which analytical reasoning was then regarded.

As with the measurement of the height of the telephone pole, Eratosthenes's method makes use of assumptions beyond that of a spherical earth. Three have been stated already: that measurements can be made with sufficient accuracy; that it is valid to apply Euclidean geometry to the process; and that distances and measuring devices (sticks or rods or strings of known lengths) are fixed quantities. A new one is that the radius of the earth is so much less than the sun–earth distance that rays of sunlight are parallel to one another when they hit the earth. This assumption is valid to a very high degree of accuracy. Eratosthenes undoubtedly accepted all of them without reservation—he may never have thought to question them. Nevertheless, they are assumptions, ones that need not, and do not, hold in all circumstances.

On the Use of Appropriate Units

The preceding assumptions are critical to the particular measurement process. In a different category are the choices of units in which to express various measured quantities, especially distances. It is standard practice to use miles or kilometers when the distances are large compared with the lengths of typical human or household objects, which are measured in inches and feet or centimeters (cm) and meters (m). Why are the latter units not used for distances on the earth's surface, or for its radius, or, especially, for astronomical distances? One answer is convenience: there is too much "bulk" when the smaller units are used, just as would be the case if you were forced to use coins rather than paper money when paying a large bill in cash.

The bulkiness of the smaller units refers to the quantity of numbers involved, as is aptly illustrated by the earth's radius. Recall that a mile is equal to 5280 feet or 63,360 inches, and a kilometer is equal to 1000 m or 100,000 cm. Using inches and centimeters as the units for the earth's radius, which from now on I will denote by D_E (the letter D stands for a distance, including that of the earth's radius, and the subscript E signifies the earth), its value in these units is equal to either 251,095,680 inches or 637,800,000 cm! The size of these numbers should make it clear that miles and km are the more appropriate units, if for no other reason than not wishing to take the time and space to write out nine digits as opposed to four. However, there is another reason for not using the preceding pairs of nine numbers, one related to the concept of *significant figures*. How many of the digits in each set of nine are needed for both accuracy and understanding, as opposed to precision? That is, how many of the nine digits are *significant*—or meaningful—in the particular context? The general answer to a significant-figures question depends entirely on the amount of inaccuracy that can be tolerated.

There have been situations in science, particularly in atomic and elementary-particle physics, where the determination of as many significant figures as possible has been the key to progress: new experimental values have led to major developments in theory, and on occasion the reverse has been true. As I will show in later Chapters, careful measurements to a sufficient number of significant figures have played essential roles in astronomy and cosmology. But, in the present context, maintaining ultimate precision is generally unnecessary: because my discussions are descriptive and not technical, no vital information will be lost by keeping only a few, rather than the entire set of nonzero digits in any large numbers.

A case in point is the value of D_E: if it is approximated either by 251,000,000 inches or 638,000,000 cm, no information vital to our purpose is omitted: the essential information resides in how many hundreds of millions there are, not in how many hundreds of thousands. The errors made by using the previous approximations are just a few hundredths of a percent, which is insignificant for our purposes. However, if one were to insist on employing the smaller units—which I do not—there is another argument behind using the approximations just cited: the mile or kilometer values of D_E are themselves approximate. Retaining the precision of all nine digits thus turns out to be an exercise in pedantry rather than in accuracy.

Even if only a few significant figures are kept, however, the total number, including the zeros, may still be bulky. The overall solution to the bulkiness problem, once only the significant figures have been retained, is to employ the power-of-ten notation. Used for both very large and very small numbers, it is described in Appendix A. In terms of this powerful notation, the value for D_E when it is expressed in the inappropriate units becomes 2.51×10^8 inches or 6.38×10^8 cm.

Copernicus, Kepler, and the Astronomical Unit

In recent years, so many new results in observational astronomy and cosmology have been publicized that it is easy to ignore how much had been learned via naked-eye astronomy. Prior to the use of telescopes, people in many parts of the world believed the physical universe to consist of the earth, the sun, the moon, the five then-known planets (Mercury, Venus, Mars, Jupiter, and Saturn), the stars, and the lesser bodies such as comets and meteors. One application of the regularity in the motions of these bodies was to create reliable calendars for various purposes, including agriculture and religion.[4] Although there may have been calendar makers in other places and at earlier times who attempted to make reliable estimates of distances to any of the bodies listed above, the efforts of the early Greeks in this regard are the best known in the West. Their most accurately measured quantity was the earth's radius D_E, and because they were unable to determine a reliable value for the earth–sun distance, only a range of possible earth–moon distances were obtained, although the lower end of this range was remarkably good.[5]

From the time of Ptolemy until the work of Nikolaus Copernicus eventually reestablished the heliocentric solar system, the accepted cosmology of pretelescopic Europe was geocentric. And, until his analysis of Tycho Brahe's (naked-eye) data led Johannes Kepler to conclude that the planets (including the earth) moved in elliptically shaped orbits, Europeans also believed that only the circle was needed to describe planetary orbits.[6] There were, therefore, two paradigmatic shifts ushered in by Copernicus and Kepler: from "geo" to "helio," and from circles to ellipses.

In a sense, Copernicus straddled these developments: he reintroduced heliocentricity but retained the concept of circular orbits (in fact, he used coplanar circles centered on the sun). From these assumptions, plus an analysis based on plane geometry and his own observational data, Copernicus deduced the relative distances between the sun and the five non-earth planets. He expressed them in terms of the unknown earth–sun distance, publishing his results in 1543. As shown later in Table 1, these relative distances were remarkably accurate, given that his was naked-eye astronomy (even more accurate naked-eye data was obtained by Tycho Brahe about 100 years later). In view of this accuracy, determination of the size of the solar system in Copernicus's model of the cosmos needed only *one* additional measurement: that of the earth–sun distance, which I will denote by D_{ES}. The need for one additional measurement holds true in the modern view of the solar system, due to Kepler and Isaac Newton.[6]

By expressing the five sun–planet distances in terms of the earth–sun distance, Copernicus exploited the fact that in an orbital system based on circles, any of the sun–planet distances can serve as the length unit for the other ones. However, planetary orbits are not circles but *ellipses*—as was known to astronomers in India as long ago as ca. 600 (cf. note 4) and rediscovered by Kepler about a thousand years later. In a circular orbit about the sun, the earth would always be at a constant distance from it, but for an elliptical orbit, the earth–sun distance is continuously changing. A new problem therefore arises: which of these varying distances should be taken as the unit for measuring all the other planet–sun distances? The solution to this problem resides in one of Kepler's three laws of planetary motion, which I consider after describing some features of ellipses.

Figure 4, which compares a circle and an ellipse, illustrates aspects of this new nonuniqueness problem. An ellipse is a geometric figure that is symmetric in both the up–down and the left–right directions. It looks like a squashed circle. Each of the two heavy points in Figure 4 is called a focus, and the ellipse itself is constructed such that the sum of the distances from the two foci to any point on its periphery is equal to a constant. With this construction, the longest straight line that can be drawn interior to the ellipse is the horizontal one of length 2a. It is the *major*

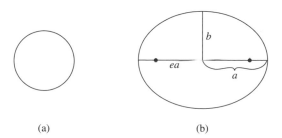

(a)   (b)

Figure 4. Comparison of (a) a circle and (b) an ellipse. The length of the semimajor axis of the ellipse is denoted *a*, its semiminor axis length is *b*, and the product of the eccentricity *e* and the semimajor axis *a*, viz., *ea*, is the distance from the center to either *focus*, each of which is specified by a heavy black dot.

axis of the ellipse. A vertical straight line of length 2*b* drawn
through the center of the ellipse is its *minor axis* (half these distances are the semimajor and the semiminor axes, *a* and *b*; only
the upper portion of the minor axis is displayed in the figure). The
amount of "squashing" is characterized by the *eccentricity, e,*
whose values range from 0 to 1: when *e* is zero, the ellipse becomes
a circle, whereas for *e* equal to 1, the ellipse collapses to a straight
line of length 2*a*. In Figure 4, the eccentricity is approximately
equal to 0.7.

The reason for detailing these properties is that the ellipse
plays such a prominent role in the three laws of planetary motion
that Kepler deduced from Tycho's wonderfully accurate, naked-eye
data. These "laws" are empirical in nature, in that they were
deduced from observational data rather than being theoretically
based. Kepler's first law states that the orbits are ellipses with the
sun located at one of the foci (not at the center). His second law
is both technical in character and not relevant to my analysis, and
I am therefore omitting it here; interested readers may look it up
in Harrison (2000) or Webb (1999). The third law is a *universal*
relation between the semimajor axis of the orbit and the corresponding period—the time it takes the planet to make a full
revolution about the sun (1 year in the case of the earth).

Because of the universality of this latter relation, Kepler (and
later Newton) chose D_{ES}, the semimajor axis of the earth's orbit,
to be the earth–sun "distance." Since the periods of the planets
were known very accurately, the third law allowed Kepler to calculate each of the five planet–sun distances with some precision;
he expressed them, of course, in units of the then unknown D_{ES}.
Now denoted the *astronomical unit,* abbreviated AU, D_{ES} sets the
scale of the solar system. In keeping with the choice of D_{ES} as the
earth–sun "distance," the other semimajor axes are each defined
as the "distance" of its planet from the sun. Are the deviations
from circularity of the orbits very large? No: the eccentricities of
most of the planetary orbits are less than 0.1, so that the sun is
much closer to the center than to the periphery of the planetary
ellipses.[7]

The values of the five planet–sun distances determined
by Copernicus and by Kepler are shown in Table 1. Those of
Copernicus are naked-eye results, whereas those of Kepler are

Table 1. Planet–Sun Distances Expressed in AU*

Planet	Copernicus's values	Kepler's values
Mercury	0.38	0.387
Venus	0.72	0.723
Earth	1.00	1.000
Mars	1.52	1.524
Jupiter	5.22	5.200
Saturn	9.17	9.531

*Values from Webb (1999).

based on his first and third laws of planetary motion (themselves deduced from naked-eye observations). The excellent agreement between them was a strong motive for accurately measuring the astronomical unit D_{ES}, since its measurement determines all the others.

Parallax

Prior to the invention of radar, a telescope was required to obtain even an estimate of D_{ES}. It took more than 200 years after the telescope's invention—probably in the early 17th century—before D_{ES} was measured with an accuracy close to that obtained using radar. The method used was parallax.

Anyone with binocular vision should easily grasp the concept of parallax, as it is the brain's intrinsic method for estimating distance. It relies on the fact that for objects not too far away, each eye sees a different image, the images being slightly displaced from one another against the background common to both. The following simple experiment shows how this works: stretch either arm to its fullest extent in front of your face, raise your thumb, and then look at it twice, first closing one eye and then the other (it is here that binocular vision enters). By carrying out this exercise, you should find that the position of your thumb moves relative to the fixed background (from right to left or left to right, depending on which eye was closed first). A similar situation arises whenever binocular vision is used to observe a not-too-distant object. In every case, with both eyes open, the brain melds the two

images into a single one; by so doing it becomes a distance esti-
mator (of course, this means of estimating distance becomes
refined by experience).

In the case of binocular vision, parallax refers to the brain's
melding of the images seen by the two eyes. In an astronomical
context, parallax refers to the observation of an object from two
different vantage points, typically from well-separated points on
the earth or from two points on the earth's orbit separated by 6
months. Figure 5 illustrates the geometry involved: the observation points are labeled 1 and 2; the object, here taken to be a point,
is labeled O; and the distance to O from the midpoint between 1
and 2 is denoted D. The light from O that reaches the observation
points 1 and 2 is represented by the dashed lines in the figure (this
light is either emitted, as from a star or galaxy, or scattered, as in
the case of a planet). The angle A between the line D and the lines
from either 1 or 2 to O is the *angle of parallax*.[8] Note that by identifying points 1 and 2 with a person's eyes and point O with his
or her thumb, this construction encompasses the binocular vision
example just described.

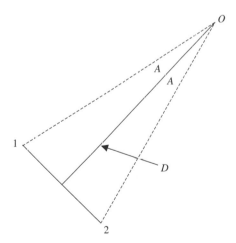

Figure 5. Illustration of the method of *parallax* (or *trigonometric parallax*)
for measuring distances. The object O is at the distance D, whose length
is to be measured. Points 1 and 2 are the locations of the two places where
the observation of O occurs; the distance between points 1 and 2, indicated
by the heavy line connecting them, is presumed known and, in the method
of *horizontal parallax*, shown in the figure, A is the angle of parallax.[8] The
method requires that A be measurable.

In the figure there are three different distances to the observation point O, namely, the distance D plus the separations between O and the two observation points. Although any one of them could qualify as *the* distance to O, the astronomical application is often to the determination of D, which will henceforth be designated as the desired *distance to be measured*. Because parallax is a known-distance/measured-angle procedure, each of its two elements must be determined beforehand. The known distance is the separation between the observation points 1 and 2, while A is the angle to be measured. In principle, the parallax angle is measured by means of observations made on the object from the two vantage points (each set of observations is made against the fixed background—the "fixed" stars[9] in the case where O is a nearby star).

There is a caveat associated with this procedure, as suggested by the phrase "in principle." It is based on the fact that as D increases, A decreases toward zero. Although a zero angle would occur only at an infinite distance, in practice D should not be so great that the angle A becomes too small to determine; that is, if O is too far away, there will be no measurable parallax. This sets a limit on the use of parallax to determine the distances to stars; correspondingly, high accuracy is required. In addition, you should bear in mind that the method of parallax involves most of the other assumptions noted previously; for example, that use of Euclidean geometry is valid.

Even taking account of the need to exercise care in measuring the angle of parallax, the procedure as just described is less straightforward than it might seem when applied to the earth–sun distance. The problem is that the object O in Figure 5 is a point, whereas the sun has an obvious size, in contrast with every other star seen from earth. Indeed, the angular widths of the sun and of the moon when it is closest to the earth are about the same— approximately 33'15"—thus allowing for spectacular lunar eclipses of the sun. The non-point-like character of the sun can be overcome, as was realized in the early 17th century, by measuring the parallax of either a planet or an asteroid as it transits the face of the sun. Combining geometry and Kepler's third law with the latter measurement allows D_{ES} to be determined.

Table 2. Numerical Values of Planet–Sun Distances and Eccentricities*

Planet	mi	km	AU	e
Mercury	35,980,000	57,910,000	0.3871	0.206
Earth	92,957,000	149,600,000	1.0000	0.017
Pluto	3,669,500,000	5,905,400,000	39.5177	0.249

*Values from Abell (1975) and Webb (1999).

After two centuries of attempts, the Scottish astronomer David Gill became the first person to measure a highly accurate value of D_{ES} using parallax. In 1888–1889, he used asteroids to determine that the AU was slightly less than 93,000,000 miles, very close to the value obtained using radar. To an excellent approximation, the current value for D_{ES} is 92,957,000 miles or 149,600,000 km. From D_{ES} one finds that the sun's radius is 696,000 km. To put this number in perspective, if the earth were an aspirin-sized sphere of diameter 0.63 cm ($^1/_4$ inch), then the sun's diameter would be 69 cm (27 inches), or about twice the size of a soccer ball!

The semimajor-axis distances plus corresponding eccentricities for several planets are given in Table 2. A planet's eccentricity is a measure of its departure from a circular orbit and can be used to evaluate both the *perihelion*, the distance of closest approach to the sun, and the *aphelion*, the farthest distance from the sun. Mercury and Pluto have the largest values of e, so that in terms of percentages, they show the greatest differences between perihelion and aphelion. That Mercury's eccentricity is so large was a plus in helping to establish the validity of Einsteinian gravity (general relativity), which I discuss in Chapter 6.

Merely because an eccentricity is very small is no reason to think that the numerical difference between the perihelion and aphelion distances will be small when expressed in miles or kilometers rather than astronomical units. An example is the earth, wherein the small value of its eccentricity translates into a perihelion–aphelion difference of about 3.1 million miles (or about 5 million km). At roughly 124 times the earth's circumference, this distance would take a 600 mph jet plane more than 200 nonstop-days to cover. Rather than simply covering this distance, and ignoring various problems such as that posed by the very high solar temperature, the same plane would need to travel for about 17.5 years in order to reach the sun itself! Furthermore, if an automobile moving at a constant speed of 60 mph were to replace the plane, each of the two preceding travel times would be increased by a factor of 10. Obviously, the planet–sun distances are nothing like ordinary terrestrial distances, yet, as you will see later, the former are miniscule compared with the distances to nearby stars, and even more so to nearby galaxies, much less to very distant ones. As big as the solar system appears, it is tiny on the cosmological scale.

The distances listed in Table 2 are a further illustration of the utility of using appropriate units, exactly as in the case of the earth's radius. Just as millions of centimeters reduce to thousands of kilometers when referring to the earth's radius, so do the millions of kilometers expressing planet–sun distances reduce to less than 100 AU when the latter distance becomes the unit of choice. Although the AU has been used as an example of the appropriate unit for expressing solar system distances, it is not especially useful when discussing the distances to stars other than the sun, even to nearby stars, the next step in our journey.

The Distances to Nearby Stars

For more than 200 years, starting with the earliest telescopic observations in the 17th century, all attempts to determine stellar distances failed: not a single parallax was detected. The best of these attempts used the earth in its orbit 6 months apart (January and June, say) for the points 1 and 2 (Figure 5), so that the distance between the two vantage points (the baseline) was 2 AU. As Galileo and many astronomers after him realized, nondetection meant that the stars were so far away that the parallax angle A of Figure 5 was simply too small to measure with the telescopes then available.

Since parallax was not observed, other methods were considered, some quite clever. One was that of Isaac Newton, who based his analysis (ca. 1672) on the fact that the brightness of Saturn and of the star Sirius (actually a double or binary star) was about the same. Using a little theory, the diameter of Saturn, the Saturn–Sun distance, and a few assumptions, he estimated that Sirius was 800,000 AU away. Converting to terrestrial units, 800,000 AU is approximately equal to 74 trillion miles or 120 trillion km; no matter which units are used, Sirius is well outside the solar system. At this distance, the parallax angle is 0.26″, much too small to be observable at that time. Just how small is 0.26″? It is, as Webb (1999) comments, about the angular width of one's thumb at a distance of 12 miles. A thumb at that distance is a trifle hard to detect with the naked eye, though nowadays it could be seen with a very good telescope. As an amusing historical footnote, Newton's estimate was only about 50% too large: the actual value is approximately 544,000 AU, a distance that is still outside the solar system.

It should be evident that the existence—or development—of the right tools is as necessary in astronomy as in repairing motorcycles or in neurosurgery, and much effort was expended improving optical telescopes in the late 18th and early 19th centuries. By the late 1830s, the necessary improvements were at hand, and by the end of 1840, parallax and the corresponding distances had been estimated for the stars Alpha Centauri, 61 Cygni, and Vega.

These stars are located in our Galaxy, the Milky Way, and are therefore all "nearby." Nearby in this case, of course, is not very close; this, coupled with instrumentation errors, led to inaccuracies in the estimates of the parallax angles. In turn, the distances obtained from them were also inaccurate, but this in no way diminishes the accomplishments of the three astronomers who finally made these first stellar distance measurements. They were T. Henderson (Alpha Centauri), F. Bessel (61 Cygni), and F. Struve (Vega). Table 3 lists the three stars, their first (estimated) parallax determinations along with their modern values, and the distances in both miles and astronomical units. The difficulty of making reliable determinations of small angles, and the errors involved, a topic always of concern in astronomy and cosmology, are clearly seen in the table. To get a feeling for these numbers, let 1 AU be represented by 0.63 cm, as with D_E (p. 23). Then the distance to

Table 3. The First Stellar Parallaxes/Distances*

Star	Est. Parallax	Modern parallax	Distance (mi)	Distance (AU)
Alpha Centauri	1.26"	0.742"	26×10^{12}	278,000
61 Cygni	0.31"	0.287"	67×10^{12}	719,000
Vega	0.265"	0.125"	150×10^{12}	1,560,000

*Values from Webb (1999).

Alpha Centauri would be 1.75 km, and that to Vega would be 9.8 km!!

It is easy to surmise, in the present era of rapid advances in microchip processor speeds or molecular biology, that once these first parallaxes and distances had been obtained, many others would have soon been measured. As is so often the case in science, reality was different: more than a hundred years later, fewer than 6000 parallaxes had been determined to within an accuracy of 0.01" (i.e., a hundredth of a second). Of the various problems that had prevented further progress, the effects of the earth's atmosphere were probably the most significant: the atmosphere is turbulent, leading to a blurring of images; it also bends (refracts) the light passing through it. The only way to eliminate these latter problems is literally to put the atmosphere behind, instead of in front of, the telescopes. That is, telescopes will do far better if they are in a satellite whose orbit is above the earth's atmosphere. Thus were the Hubble Space Telescope (HST) and the Hipparchos satellite missions initially conceived.

The HST has produced a very large number of images, the spectacular beauty of which has far surpassed expectations; many of them can be found on the HST Web site, as well as in popular astronomy magazines. But, because the HST was not designed exclusively for parallax measurements—it is serving many other functions—the determination of both stellar positions and parallax became the task of the Hipparchos mission and its Tycho experiment. Positional information accurate to 1 millisecond (0.001") has been obtained for more than 100,000 stars, and more than 1 million positions have been determined to an accuracy of

0.01". The distances to more than 22,000 stars were measured to an accuracy of about 10%, the farthest being about 1.8×10^{15} (1.8 quadrillion) km away!

The latter number is a distance about ten times that of Vega from the sun. It is not only huge, it is essentially the maximum distance for which parallax is a viable method: for larger distances, the associated error becomes unacceptably large. Control of errors, as I have noted, is a crucial element in assessing the reliability of distance measurement methods. But before I consider how parallax uncertainties affect the measurement of the distances, I will introduce the ultimate set of units in which astronomers and cosmologists express distances, since one of them, namely the *parsec*, is defined in terms of a specific angle of parallax.

Astronomical and Cosmological Distance Units

The new units are based on the fact that light travels at a very great but finite speed. This speed, denoted by the letter c, is 299,792,438 meters per second (m/sec, where "sec" is the abbreviation for seconds), or about 3×10^5 km/sec—equivalently, about 186,000 mi/sec. On multiplying c by a relevant time (expressed in seconds), light *speed* is turned into the *distance* light has traveled during that particular time. For example, light travels approximately 300,000 km (or about 186,000 miles) in one second, a distance that is a *light second* in length. A light second is not a very large distance on the scale of interest to astronomers, cosmologists, and us, but the *light year* is. The light year (abbreviated *ly*), the distance light travels in 1 year, is the smallest of the new units. It is a unit you may have seen in articles on recent developments in astronomy or cosmology. How many kilometers or miles are contained in a light year? The answer is obtained by first finding the number of seconds in 1 year (about 31.5 million) and then multiplying this number by the numerical value of c. The result: a light year is approximately equal to 9.45×10^{12} km (or 5.88×10^{12}

Table 4. Stellar and Galactic Distance Units

These units, all based on the speed of light, are

The **light year** (**ly**) is the distance light travels in 1 year
The **parsec** (**pc**) = 3.26 ly
The **megaparsec** (**Mpc**) = 10^6 pc = 3.26×10^6 ly

The speed of light is denoted c; its value is

$\quad c = 2.9979 \times 10^5$ km/sec $\cong 3 \times 10^5$ km/sec
or
$\quad c = 1.8627 \times 10^5$ mi/sec $\cong 1.86 \times 10^5$ mi/sec

In one second, light travels 299,792 km or 186,280 mi. To obtain a distance based on c, multiply c by the elapsed time; to get the time required to travel a distance, divide it by c. For example,

The sun is about 8.3 light minutes away
The moon is about 1.3 light seconds away
The earth's circumference is about 1/7 light second.

There are approximately 31.5 million seconds in a year, so on multiplying this number by the above values, one finds

1 ly $\cong 9.45 \times 10^{12}$ km (5.88×10^{12} mi)
1 pc $\cong 30.8 \times 10^{12}$ km
1 Mpc $\cong 30.8 \times 10^{18}$ km (!)

---

miles). Referring back to Table 3, it is clear that the ly can play the same role for stellar distances that the AU does for solar system distances (Table 2).

The next unit in this new set is the *parsec*, denoted *pc*. It is the distance at which the angle of parallax is equal to 1 second, and that turns out to be 3.26 ly, or about 30 quadrillion km. Although both the light year and the parsec are big, neither is quite large enough to satisfy the needs arising in cosmology, and so one more unit has been added to them. It is the *megaparsec*, abbreviated *Mpc*, equal to a million parsecs, or about 30×10^{18} km.

Table 4 summarizes these and a few other results. It also introduces the symbol $\cong$, which represents the phrase "approximately equal to," and is an equals sign with a squiggle over it. You will encounter the new symbol from time to time in the remainder of the book.

The light year is more than a handy unit: by expressing a distance in light years, you immediately know how many *years* it will take light to travel that distance. Hence, the information carried by radiation that has traveled an arbitrary distance of, say, N light years tells us about the conditions of the emitter N years ago. This is the means by which the archaeology of the Universe is quantified. If an object a billion light years away emits radiation, the radiation not only takes a billion years to reach the earth, it also carries information that was current a billion years ago rather than now. Radiation is cosmology's archaeological tool, allowing for exploration of the Universe as it was in the past. The concept of the "current Universe"—as it is now—is operationally meaningful only in our immediate vicinity: the farther away an emitter is, the further back in time are we seeing it. Moreover, the information deduced will always be about past behavior: to infer how objects N light years away are behaving *now*, observers will need to wait until N years have passed before beginning their observations. The light from distant objects is ancient: our now is always the emitter's then.

As a first use of the new units, I have the restated in Table 5 the results listed in Table 3, with the distances now expressed in light years and parsecs. Also given are the times in the past when the light detected on earth was emitted by the three stars. As a casual definition, the concept of "nearby stars" can now be interpreted as those whose light was emitted not too long ago, within, say, the past hundred or perhaps the past thousand years. Although these times are somewhat arbitrary, nearby stars are restricted to lie within the Milky Way Galaxy, some properties of which I discuss in Chapter 5.

The maximum distance for which parallax yields reliable results is about 90 parsecs, or roughly 300 light years. Inaccuracies in measuring the parallax angle A, leading to unacceptable distance uncertainties, underlie the existence of this limit. As I noted

Table 5. Stellar Distances in Light Years and Parsecs

Star	Distance (ly)	Distance (pc)	When its light was emitted
Alpha Centauri	4.39	1.346	4.39 years ago
61 Cygni	11.35	3.48	11.35 years ago
Vega	24.30	7.45	24.30 years ago

previously, the larger the distance D in Figure 5, the smaller is A, and therefore the more difficult it is to measure A accurately. The relation between D and A is especially simple if A is expressed in seconds and D is given in parsecs, for in this case, they are the inverses of one another. That is, D is equal to 1 over A, and conversely, A is equal to 1 over D. In place of the foregoing written statements, one can re-express them with an equivalent pair of equations, whose elements are D (stated in pc), A (expressed in sec), and the equals sign, =, as follows: $D(\text{pc}) = 1/A(\text{sec})$ and $A(\text{sec}) = 1/D(\text{pc})$. From the first of these relations, one finds that if the parallax angle is 1 second, then $A = 1$, in which case $D = 1/1 = 1$, or 1 parsec, as this formula requires D to be expressed in parsecs.

Some Comments on Errors

I have pointed out several times the occurrence of errors and the need to control them. Errors may arise in a variety of ways, and after a short discussion of how they can occur in a simple—but still real-world—length measurement, I shall apply the preceding analysis to it.

Errors will be present whenever a quantity needs to be known to an accuracy better than is afforded by the calibration of the measuring instrument, be it a scale for determining weight, a ruler for measuring length, etc. In this situation, the measurer (the analogue of the "observer") must *interpolate* between adjacent markings on the finest scale of the instrument. The measurer will then end up with an *interpolated* number, rather than an exact one, and the measured value of the quantity will then have an uncertainty or error attached to it.

The uncertainty could be by an amount equal to half the distance between the fine scale markings on the instrument. To be specific, imagine that the measuring instrument is a meter stick whose finest rulings are in centimeters. Required is a length accurate to a tenth of a centimeter. Suppose that after placing one end of the object to be measured at the 0 mark of the meter stick, the other end is found to lie between the marks for 45 and 46 cm. For the sake of this example, it is assumed that the best the measurer can do is estimate the position to be between 45.4 cm and 45.6 cm. She could then state the desired length as 45.5 ± 0.1 cm, where the quantity ± 0.1 cm means that 0.1 is to be added to and then subtracted from 45.5 to get a low value of 45.4 cm and a high value of 45.6 cm. The uncertainty (or the error) in the measured value of 45.5 cm is 0.1 cm.

The hypothetical situation just outlined contains the essential elements that arise in making measurements on very small parallax angles. Consider first a measurement of A yielding a value of 0.01 ± 0.003″. The implication of this result is that D ranges between 77 pc and 143 pc, the latter value being about twice the former. The large measurement error (uncertainty) of 30% in A leads to a corresponding uncertainty of 30% in the distance D, which for most purposes is much larger than is desirable. On the other hand, if the object were closer and the measured uncertainty reduced in value, for instance, if A were changed to 0.1 ± 0.002″, then the range of D would be from 9.8 pc to 10.2 pc (10 ± 0.2 pc). The inaccuracy in distance is now much more acceptable. Unfortunately, 10 pc is not very far away on an astronomical scale, so that this reduction in uncertainty has not gone hand in hand with a further reach in distance. The problem that this example illustrates is typical: the farther away the object, the bigger the error.

Because the upper limit on distances yielded by parallax measurements does not reach beyond the Galaxy, but almost all the objects in the Universe are extra-Galactic, methods for determining distances other than parallax are needed. Many exist (otherwise the science of cosmology would have no observational component); all involve concepts that are used in the theoretical description of stars, radiation, and the Universe. One is the concept of *mass*; another is its related concept, *density*. Among other properties, mass determines a star's evolutionary end-stage; while whether the Universe is open or closed, as well as flat or curved, is partly determined by the density of matter in it. Mass is a fundamental ingredient in Newton's laws of motion, especially in his law of gravity, which provides the theoretical foundation for Kepler's three laws. Mass and density are the last items discussed in this Chapter, which is followed by chapters on light and radiation and on stars.

Mass

Mass is a measure of the quantity of matter in a body, such as an atom, a person, a car, a star, a galaxy, and so forth, and may be thought of as an entity abstracted from the more tangible notion of weight. On the earth's surface, the weight of an object is equal to the product of its mass and a certain number denoted g, which is a measure of the gravitational force exerted by the earth on the object. Conversely, an object's mass is equal to its weight divided by g. (The value of g on the earth[10] is different than its value on the surfaces of the sun or the moon: the sun's value of g is larger than the earth's, whereas the earth's is larger than the moon's.) Without necessarily knowing exactly what mass is, most people know that in Einstein's famous equation $E = Mc^2$, M is the mass of the object whose energy is E. The name is familiar even if the details of the concept may not be.

The primary unit of mass I use in this book is the kilogram, denoted kg. Masses, like the sizes of the objects found in the Universe, cover an enormous range of values. To give an illustration of this range, and confining the set of bodies to those occurring in the solar system, I have listed in Table 6 a selection of masses (abbreviated M), starting with the proton and ending with the sun.

The sun's mass is the unit in which stellar and galactic masses are often expressed, so it becomes an analogue of the light year and parsec. That the mass of a person, apart from a few factors of ten, falls approximately midway between a proton's mass and the earth's mass, is intriguing but is not especially significant from

Table 6. Masses of Various Bodies

Body	Mass (M)
Proton	$M_{proton} = 1.67 \times 10^{-27} \, kg$
Person	$M_{person} \cong 50 - 75 \, kg \cong 50 - 75 \times 10^{27} \, M_{proton}$
Earth	$M_{Earth} = 5.98 \times 10^{24} \, kg = 3.6 \times 10^{51} \, M_{proton}$
Jupiter	$M_{Jupiter} = 1.9 \times 10^{27} \, kg$
Sun	$M_{Sun} = 2 \times 10^{30} \, kg \cong 1000 \times M_{Jupiter}$
	$\quad = 3.33 \times 10^5 \, M_{Earth}$
	$\quad = 1.25 \times 10^{57} \, M_{proton}$

a cosmological perspective. The reader may wish to consider its possible implications for other areas of human activity or thought. That Jupiter's mass is about a thousandth of the sun's mass also lacks any significance cosmologically: as will be seen in Chapter 4, it would need to be at least 80 times larger for this to be the case. Nor is Jupiter a *failed star*, which for reasons I discuss in Chapter 9, is a designation applied not to planets but to brown dwarfs.

The foregoing comments may suggest that mass is simply a passive attribute of matter. Far from it! Mass is the active generator of gravitational effects. In Einsteinian gravity, it warps space in a way that causes both matter and light to deviate from the paths they would take in the absence of mass, as I discuss in Chapter 6. For Newton's theory of gravity—the reigning one until Einstein's—it is the generator of the gravitational force. In what was a triumph of the intellect far surpassing that of Eratosthenes and every one else until Einstein 250 years later, Newton not only formulated a mathematical theory of the dynamics of moving bodies, but he also specified the detailed mathematical form of the gravitational force between any two masses.

Newton's gravitational force is proportional to the product of the two masses divided by the square of the distance separating them. The idea of an *inverse square force* was not original with him: Kepler had speculated that such a dependence on the separation might describe the force exerted by the sun on the planets, and some of Newton's contemporaries had considered it. However, Newton's genius permitted him to do what none of the others could. Along with Gottfried Leibniz in Germany, but independently of him, Newton invented the branch of mathematics known as *calculus* and, sometime in the period 1664–1668, used it to *calculate* the orbit of a planet acted on by his gravitational force.

The answer was an ellipse! He later proved that Kepler's other two empirical laws were also a consequence of the inverse square force. These and his many other findings revolutionized science, giving rise in part to the concept of a completely deterministic universe. And, even though Einstein's version of gravity displaced Newton's, the latter is a valid approximation except when the force is strong, as in the case of a black hole, or when extraordinary precision is required (as in GPS, Chapter 6).

Newtonian gravity certainly applies in the solar system. The quantity g that connects mass and weight is obtained from Newton's gravitational force. Both the sun and the moon exert inverse square gravitational forces on the earth, but that of the sun is the stronger despite its being much farther from the earth than the moon: the much greater value of the sun's mass more than compensates for its having the larger separation from the earth. Newtonian gravity also works quite well in describing much of the Universe: Einsteinian gravity need not be brought into play in accounting for a variety of stellar and galactic properties, as will be seen later.

At the other end of the mass scale, gravity has essentially no effect on the behavior of atoms, molecules, nuclei, or electrons under the ordinary conditions encountered on the earth or on the surfaces of most stars, essentially because of the miniscule masses involved. In Chapter 4 I will quantify just how insignificant is the force of gravity between two protons, as compared with the electrical force between them.

Density

For cosmology, density is as central a concept as mass. Denoted d, density is equal to a body's mass divided by its volume and is expressed in units of kg/m^3, where m is the abbreviation for meter ($= 10^{-3}$ km). Although kg/m^3 is the standard unit, the density of water often replaces it as the unit in which the density of other objects is expressed. This choice may not be surprising to you, as both the human body and the earth's surface consist mostly of water (about 70%). In addition to mass density, one can also introduce energy density (energy per volume), but because energy and mass are related by Einstein's equation, I use the symbol d for each: the context will indicate which density is meant.

The presence of m^3 in the denominator makes density a non-intrinsic property of objects: it can be made larger or smaller by changing the size of the volume. Nonetheless, it is sometimes helpful to be aware of certain densities, especially when the volume is taken to be a cubic meter, since they play a role in ordi-

Table 7. Densities of Various Bodies

Body	Density (d)
Water	$d_{water} = 10^3 \, kg/m^3$
Lead	$d_{lead} = 11.3 \, d_{water}$
Earth	$d_{Earth} = M_{Earth}/\text{Earth's volume} = 5.5 \, d_{water}$
Sun	$d_{Sun} = M_{Sun}/\text{Sun's volume} \cong 1.41 \, d_{water}$
White dwarf star	$d_{White\ dwarf\ star} \cong 10^{10} \, d_{Sun}$
Neutron star	$d_{Neutron\ star} \cong 10^{15} \, d_{Sun}$
Luminous matter	$d_{Lum\ matt} \cong 3.2 \times 10^{-29} \, kg/m^3$

N.B.: d_{Earth} and d_{Sun} are average values.

nary situations as well as in some cosmological ones. Table 7 lists densities for a variety of bodies.

That lead is used as shielding against radiation is a consequence of its high density, but what may be surprising is that the average—or *mean*—density of the sun is so close to that of water! This reflects the fact that the surface of the sun is more like gossamer, whereas at the center its density is far greater than that of lead. (Averages alone cannot fully describe a complex situation.) White dwarfs and neutron stars are two of the end points of stellar evolution that arise from the collapse of stellar matter into volumes that are very much smaller than that of the original star. It is the occurrence of the relatively small volumes in their denominators that leads to such large values of the corresponding densities. These exotic objects as well as black holes are described in Chapter 4.

Perhaps the most astounding item in Table 7 is the tiny value for the density of luminous matter, which consists mainly of stars. The luminous part of the Universe is mostly empty space—just like the Universe itself, as you will see in Chapter 6. The value of $d_{Lum\ Matt}$ corresponds to about two protons per $100 \, m^3$. It is the least certain of the densities listed in the Table,[11] made difficult to estimate partly because there is so much matter in galaxies that doesn't shine. This nonluminous, or *dark*, matter is five to six times more abundant than all of the ordinary matter in the Universe, a feature I shall discuss in Chapter 7.

3. Light, Radiation, and Quanta

For thousands of years, people observed stars and planets solely by that marvelous instrument, the unaided human eye. And, despite the various forms of electromagnetic radiation populating the environment, such as microwaves, radio waves, ultraviolet and infrared waves, and so forth, the human eye still responds only to the visible portion of the electromagnetic spectrum of radiation. That visible portion is, of course, what we call *light;* the remainder is usually referred to as *radiation.* Human eyes are sensitive to only two characteristics of light: its *wavelength* and its *intensity.* For the moment, you may think of wavelength as referencing the color of the light; I will present a more technical definition shortly. Intensity means the amount of light and is a measure of intrinsic brightness: a 1000-W bulb is more intense than a 100-W bulb; the sun is far more intense than either. It is usually easier to see a source of greater intensity than it is a weaker one, but just as too weak a source will not lead to seeing, too strong a source cannot be viewed for too long.

When you see something, light has entered your eye either directly from its source or indirectly, when it is scattered by an object, yet it was once thought that people saw by means of light emitted from the eyes. Furthermore, it was also believed that the rainbow-like colors seen when sunlight shone on a glass prism were intrinsic to the prism. It wasn't until Isaac Newton performed experiments around 1662 that sunlight itself was understood to be composed of colors.

Newton carried out two experiments, whose main ingredients were a darkened room, sunlight passing through a pinhole in a screen or shade covering a window, a triangular prism, and a second screen. They are schematically illustrated in Figure 6. In the first experiment, the beam of sunlight passed through a single prism and was then transmitted to the second screen, where colors ranging from red to violet were seen. This verified previously known results.

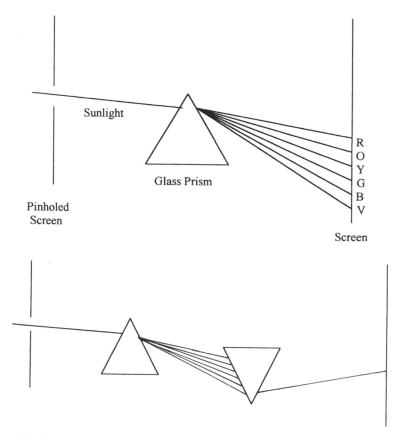

Figure 6. Illustration of Newton's experiments on the decomposition of sunlight (white light) into its constituent colors. In the upper portion of the figure, a beam of sunlight passes through a slit in a screen and then onto a triangular glass prism. Its passage through the prism decomposes the light into its rainbow colors, which are indicated on the screen to the far right by the rainbow colors red (R), orange (O), yellow (Y), green (G), blue (B), and violet (V). In the lower part of the figure, a second prism, inverted with respect to the first, is interposed between it and the screen. The effect of the second prism is to recombine the colors into a single beam of white light, also indicated on the screen.

The crucial experiment was the second one, in which a second triangular prism, inverted relative to the first, was added.[1] All the colored light from the first prism fell on the second one, passed through it, and was directed onto the screen in the still darkened room. The light that now hit the screen was again white! Prism 2 had eliminated the colors, an event consistent only with the colors being intrinsic to the light, not to the prism.

With this experiment, Newton had discovered that white light is composed of the same spectrum of colors as seen in a rainbow. The role of the prism was either to decompose visible light into its spectrum of colors or to recombine the spectrum into white light.

In addition to discovering the spectral decomposition of white light, Newton also speculated on its nature, concluding that it consisted of little particles he called *corpuscles*. He then used this conclusion to explain the presence of different colors in sunlight. Although his corpuscular theory was wrong in detail, the concept itself was correct: light, and indeed all electromagnetic radiation, is composed of massless particles called *photons*, which are discrete bundles or *quanta* of electromagnetic energy. It took well over 200 years before the photon nature of light was hypothesized, much less accepted by the scientific community, in part because the prevailing paradigm had been that of waves. How waves came to replace corpuscles, with photons then supplanting waves, is one of the most intriguing stories in the history of science, portions of which I shall recount below. It begins with another look at Figure 6.

This figure is highly schematic, in that the result produced by the first prism is shown as six separate lines, corresponding to the six rainbow colors red, orange, yellow, green, blue, and violet; these are the hues commonly understood to make up the visible spectrum. In reality, though, the colors are not sharply demarcated. Instead, they transform smoothly from one to the next, in a continuous manner (Plate 1). In explaining this continuous spectrum by means of his corpuscular postulate, Newton created a paradigm that lasted for almost 150 years. However, he overlooked a feature in the spectrum that later became the key to unlocking the mystery of stellar atmospheres and the basis for demonstrating the expansion of the Universe.

The item Newton failed to notice was the presence of dark lines in an otherwise continuous spectrum of colors. One hundred forty years later, when William Wollaston repeated Newton's experiment, he found seven dark lines in the continuous spectrum, which he incorrectly interpreted as natural boundaries between the colors Furthermore, Newton would not have been able to explain them correctly either, even if he *had* seen them. They are shown in Plate 1, which demonstrates the continuous nature of the solar spectrum. I will have more to say about this color plate later.

Because of their future importance, the discovery of the dark lines may be considered the forerunner of several revolutions in the sciences of light, radiation, and the structure of matter. These paradigm-changing events began with the 1804 experiments of Thomas Young, done just 2 years after the experiments of Wollaston. Although Young also used a darkened room and two screens, he made a crucial modification to the screen covering the window: he added a second pinhole to it. The far screen served the same purpose as in 1662: to display the light transmitted through the two pinholes.

When either of the pinholes was covered, the far screen showed a bright area surrounded by shadow. But when both holes were open, so that sunlight passed through each, he observed a pattern of alternating dark and light areas on the far screen. This pattern of lesser and greater intensities of light on the second screen is an *interference* phenomenon. It cannot be produced by particles striking the screen but is a natural consequence if *waves* are hitting it. The observation of interference rendered Newton's corpuscular theory untenable, and ultimately led to a paradigm shift, from light as a particle to light as a wave.

The statement of the new paradigm may well raise questions in your mind: a particle is understood to mean a tiny mass, but what is a wave, and how or why can light be wave-like (if not actually a wave)? Our environment contains many visible examples of waves, among them flags flapping in the wind; ripples spreading out in circles on the surface of a pond when a stone is dropped into it; ocean swells, possibly leading to the huge breakers that form a surfer's paradise; arms signaling hello or goodbye; the sequential standing up and sitting down of blocks of spectators at sports stadiums (aptly called "The Wave"). Less visible are the vibrations of a violin string when it is bowed; invisible is the back-and-forth motion of air molecules as they conduct sound. None of these examples define a wave, nor is it clear that any are applicable to light. The nature of waves remains to be elucidated, and this is the subject I turn to next.

Properties of Waves

The feature common to each of the foregoing examples is the motion of a medium of some sort (flag, water, arm, people, string, air) that does the actual moving. A wave is defined as a *periodic* or repetitive disturbance in a medium. Waves obey one or another type of *wave equation*, mathematical entities that were crucial for establishing the wave-like nature of electromagnetic radiation. Many types of waves spread or *propagate*; these are called *traveling* waves. In contrast are *standing* waves, such as those on a stringed instrument, wherein the string vibrates perpendicularly to its length. There is no movement at the fixed ends, and the waves are confined to the string. The string's motion disturbs the air molecules, which oscillate back and forth, each molecule disturbing its neighbors; the motion of the disturbed molecules carries energy, which eventually tickles—or hammers—our eardrums, thus allowing us to hear.

To help understand the properties of waves, scientists usually deal with simple examples, assuming the ideal condition of no attenuation due to friction. The simplest example is a *sine* wave, which travels in a frictionless medium and propagates over all space. Figure 7 shows a finite portion of the wave. The curve represents the disturbed medium, enough of which is present to display the periodicity of the wave. When undisturbed, the medium lies flat, and this is depicted in the figure by the horizontal straight line. Because the wave is perpendicular to the undisturbed medium in which it propagates, it is said to be

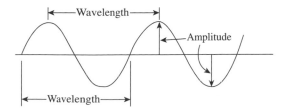

Figure 7. Schematic representation of a portion of a *sine* wave. Shown are the wavelength, which is the distance between adjacent portions of identical character (two adjacent maxima and two adjacent *nodes*, or zeros), and the amplitude, which is the maximum departure from the undisturbed position.

transverse (the waves of electromagnetic radiation, of which light is an example, are transverse).

Two features of a wave are identified in the figure: its *amplitude* and its *wavelength*. Amplitude is the maximum departure from the undisturbed position, occurring both above and below the horizontal line. The intensity of a wave—the energy it carries, or its brightness in the case of light—is proportional to the square of its amplitude. Wavelength is defined to be the minimum spatial distance over which the wave form repeats; two instances are seen in the figure: these are the distances between two sequential maxima and two sequential zeros or nondisturbances. That a wave is periodic means that the same repetition occurs more than once (it occurs an infinite number of times in the case of a sine wave).

Wavelength is usually referred to by a symbol. The one I use in this book is λ, the lowercase Greek letter lambda, pronounced "lamduh." Of equal importance to the wavelength of a wave are its *frequency* and the *speed* with which it propagates in the medium. Frequency, denoted by the letter f, refers to the number of times per second the repetition occurs. Musicians know it as *pitch*, and someone with perfect pitch knows and can hum or sing the note corresponding to a specific frequency. Frequency is also familiar to people who listen to AM and/or FM radio, where the call letters of the station are typically accompanied by a statement of the frequency on which the station broadcasts.

The propagation speed, which I denote by v_{prop}, varies with both the type of wave and the medium in which it travels. At normal temperature and pressure, the speed of sound in air at sea level is about 344 km/sec or 1100 feet/sec, whereas in water that speed increases dramatically to about 1400 km/sec. For electromagnetic radiation, which travels at the speed of light, v_{prop} is replaced by the special symbol c, introduced in the preceding chapter.

Propagation speed, frequency, and wavelength are linked by an exquisite relation. It states that the speed of propagation is equal to the product of the frequency and the wavelength. Written out as an equation involving the symbols, this relation becomes $v_{prop} = f \times \lambda$. Because this is one equation relating three quantities, only two of them are independent: if, as is usual, v_{prop} is known, then measuring or specifying the value of either member of the right-hand side of the equation automatically determines the value of the other one. In the case of electromagnetic radiation, either the frequency or the wavelength uniquely specifies the radiation. Radio stations are defined by the frequencies at which they broadcast, but nothing more than familiarity would be lost if the wavelengths were specified instead.

Frequencies or wavelengths are normally measured when the emitter of the waves is at rest, but these quantities change if the emitter or observer is moving. The physical phenomenon of changed pitch or wavelength is known as the Doppler effect, named for the Austrian scientist who derived the mathematical formula relating the new and old values to the speed at which the emitter or observer is moving. If you have heard a train's whistle or the siren of an ambulance, fire truck, or police car, then you have likely encountered the Doppler effect: the pitch increases (and the wavelength decreases) if the emitter is approaching, whereas the opposite is true if the emitter is receding. The latter case is the norm in cosmology: most galaxies are receding from earth; only a few nearby ones are approaching.

I will concentrate here on wavelengths and a source of waves *receding* from the observer. In this case, the observed wavelength will be increased compared with its value when the emitter is stationary. As you will see shortly, an increase in wavelength for light means a shift toward the red end of the visible spectrum. This phenomenon has led to the increases in wavelength for all electromagnetic radiation being termed *redshifts*, a terminology used exclusively for this case. Redshifts are one of the keys used to decipher the Universe and so play a major role in cosmology.

The cosmologically interesting version of Doppler's formula is an approximation to it that Edwin Hubble used in 1929 to infer that the Universe is expanding. The approximation equates two ratios, one of which is V/c, where V is the recession speed and c is the speed of light. The numerator in the other ratio is the difference between the observed (increased) wavelength and the original one, and the denominator is the original wavelength. It is easily expressed as a formula by using the wavelength symbol λ. By letting λ_{obs} stand for the *observed* wavelength and λ_{emit} for the original or *emitted* wavelength, the second ratio becomes

$[(\lambda_{obs} - \lambda_{emit})/\lambda_{emit}]$, a quantity usually denoted by z and referred to as the *redshift parameter*. Equating the two ratios yields the approximate form of Doppler's formula:

$$[(\lambda_{obs} - \lambda_{emit})/\lambda_{emit}] \cong V/c. \qquad (1)$$

As described in Chapter 5, Edwin Hubble used this approximation to determine the speeds V with which galaxies are receding from the earth, from which he inferred that the Universe is expanding. Relation (1) is valid as long as the ratio V/c is somewhat less than $0.2 = 1/5$, otherwise the exact formula must be used, although it is not needed in the present context.[2]

Approximation (1) involves wavelengths and can equally well be expressed in terms of frequencies. While astronomers measure wavelengths, frequencies are identified in the case of sound waves. Were musicians to play their stringed instruments on a moving platform, listeners could detect the changes in frequencies—at least in principle. Just such a method—musicians on a moving platform—was used when the Dutch scientist Christopher Buijs-Ballot successfully tested Doppler's formula in 1845,[3] although his procedure is not the pitch-altering one normally used by contemporary musicians! Instead, players of stringed instruments change pitch by pressing on one or more strings. This method works because the frequency of a string fixed at two ends is inversely proportional to the string's length. By pressing on a string, the length available for vibrating decreases to a smaller effective length, the inverse of the smaller effective length increases, and the pitch therefore rises.

Electromagnetic Waves

The wave-like nature of all electromagnetic radiation—and not only light—was deduced by the Scottish theorist James Clerk Maxwell in the late 19th century, a deduction based on his mathematical theory of electromagnetic phenomena. Although Maxwell built on the quantitative theoretical and experimental work of many others, qualitative electrical effects had been known for millennia. Plato, in the dialogue *Timaeus*, refers to the marvelous properties of amber (it attracts hair when rubbed with fur), and in the 1500s the English physician William Gilbert coined the word *electric* for this type of phenomenon: "electron" is the Greek word for amber. The American scientist and diplomat Benjamin Franklin introduced the terminology positive and negative electricity and called the amount of electricity in a body its *charge*.

The first of the quantitative results was obtained by Charles Coulomb, a French scientist who had been investigating the force between stationary charges. In 1784, he determined the dependence of this force on the separation of the charges as well as on their strength and signs. Only a few decades later, moving charges in the form of electric currents were found to give rise to magnetic effects, following which the mathematical descriptions of the forces between a charge and a current and between pairs of currents were obtained. Maxwell brought the theory of electromagnetic phenomena on the macroscopic scale to scientific closure in 1865 by codifying both older and his own theoretical studies into a unified mathematical framework consisting of four equations, which soon became known as Maxwell's equations. Combining them produced a wave equation, from which he predicted that moving charges would generate electromagnetic radiation in the form of traveling waves whose speed was equal to that of light. The wavelengths of these waves were unrestricted, covering all of what is now referred to as the electromagnetic spectrum of radiation.

Maxwell's prediction was verified in 1887 by the experimental work of the German physicist Heinrich Hertz, who not only generated radiation with macroscopic wavelengths but also showed that their speed was indeed that of light. Electromagnetic radiation was therefore conclusively established as wave-like; 14 years later the Italian engineer Guglielmo Marconi sent radio waves across the Atlantic Ocean, thereby laying the foundation for the broadcast society in which we live.[a]

---

[a] This brief summary scarcely does justice to one of the great triumphs of 19th century science, and readers interested in the history of the subject are recommended to the references listed in endnote 4 to this chapter and to the further citations contained therein, especially those of Weinberg (1983).

Table 8. The Spectrum of Electromagnetic Radiation

Segment	Wavelength (m)	Frequency (Hz)
Radio	Greater than 1	Less than 10^8
Microwave	1 to 10^{-4}	5×10^8 to 10^{12}
Infrared	10^{-3} to 10^{-6}	5×10^{11} to 5×10^{14}
Ultraviolet	5×10^{-6} to 5×10^{-9}	10^{15} to 10^{17}
X-ray	10^{-8} to 10^{-13}	10^{16} to 10^{21}
Gamma ray	Less than 10^{-10}	Greater than 10^{18}

Table 8 summarizes the main segments of the electromagnetic spectrum of radiation, with wavelengths expressed in meters and frequencies expressed in Hz. The latter is an abbreviation for *Hertz*, the nomenclature used for cycles per second, in honor of the physicist. The frequency range for FM radio is 88 MHz to about 108 MHz, whereas that for AM radio is 540 kHz to 1600 kHz, where mega (M) = 10^6 and kilo (k) = 10^3. These frequencies correspond to fairly long wavelengths, much greater than those of the ultraviolet radiation that produces sun tans (and possibly skin cancer) in lighter-skinned people.

Table 8 does not identify the range of visible light, a range so small that an expanded scale is needed to portray it. The upper portion of Figure 8 shows that scale, which is indeed miniscule compared with the enormous range of wavelengths over which the electromagnetic spectrum occurs. Notice that the longest wavelengths are at the red end of the visible spectrum, so that if an emitter of visible radiation of any color other than red were receding, the observed wavelength would be redshifted, as I noted above. The major segments of the electromagnetic spectrum (plus a few others) are restated in more graphic terms in the lower part of the figure, with the overlaps being historical rather than deliberate.

Since the tiny visible range is bracketed by the infrared and the ultraviolet, you may have wondered why we see only in the visible and not in either of these other two segments. Human vision is limited in this way because absorption by the earth's atmosphere prevents most of the infrared and ultraviolet radiation from reaching the earth's surface. Hence, as life evolved on earth, the successful adaptations were those that developed visual

Human eyes are sensitive only to the tiny *visible* range, shown immediately below:

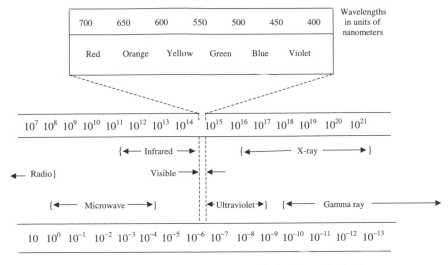

Figure 8. The spectrum of electromagnetic radiation. The lower part of the figure depicts the principal portion of the spectrum. It is bounded above by the frequency lines, containing the numbers 10^7 through 10^{21} (units of Hz), and below by the wavelength lines, containing the numbers 10 through 10^{-13} (units of meters). Shown as well are the ranges over frequency and wavelength of the various segments such as infrared, X-ray, etc. Since the visible segment, sandwiched between the vertical dashed lines, is too small to display the colors it contains, an exploded view is presented in the upper part of the figure, where the rainbow colors and their wavelengths are identified.

systems that responded to the visible portions of the sun's spectrum.[5] As shown in the next chapter, the sun's maximum intensity is in the green, with a slight falloff in intensity at neighboring visible wavelengths. This gives rise to a *yellow* sun, with the resulting radiation being seen on earth as white light.

Wavelengths and Spectra

The wavelengths of a few of the colors seen in the sun's spectrum were first measured by Thomas Young in his 1804 experiment; soon afterwards, many scientists began to measure wavelengths.

Among them was the German scientist Josef Fraunhofer, who was one of the best prism makers. By 1814, he had found hundreds of dark lines in the sun's spectrum, measuring the wavelengths of more than half of them. Some of these lines are shown in Plate 1, superimposed on the solar spectrum. They are still referred to as *Fraunhofer dark lines*, yet Fraunhofer, like Wollaston, did not understand their origin, despite his also having discovered that the wavelengths of two of the dark lines were identical to those of two *bright* lines in the spectrum of light from a sodium lamp.

This latter finding was a pivotal result. It took only a relatively short time for other scientists to learn that when heated, certain materials emit radiation—known as bright lines—whose wavelengths are specific to the emitter,[6] and that if this radiation is then passed through gaseous forms of the material, the bright lines will either diminish or vanish. This phenomenon occurs because an object absorbs and emits radiation at precisely the same set of wavelengths. Dark lines in continuous spectra therefore correspond to a diminishing or an absence of radiation at that wavelength, due to its absorption by exactly the same kind of object that initially emitted it. The collection of wavelengths emitted by an object—its radiation *spectrum*—is equivalent to fingerprints or DNA: it uniquely identifies the object, and if a spectroscopist (a measurer of spectra) is given a portion of a spectrum, he or she should be able to specify the emitter.

The accurate determination of wavelengths requires sophisticated instruments, and those developed by 19th century scientists were superb. Known as spectrographs, they eventually became indispensable members of the astronomer's arsenal. Probably the outstanding scientist among the 19th century spectroscopists was the German physicist Gustav Kirchhoff. Starting in 1859 he began looking for, and then found, terrestrial elements in the sun's outer layer (the photosphere), including sodium, calcium, copper, and zinc.

I wish I could have announced that last sentence with blaring trumpets and fireworks! Kirchhoff's discovery is one of the most significant events in science, if not in all of recorded history. Wherever spectra have been measured—from stars as well as from interstellar gas and dust—only terrestrial matter has been found.

These results lead to the stunning conclusion that apart from dark matter and dark energy (entities I will consider in Chapters 7 and 9), *the Universe contains only the same material that is found on earth*. This, in turn, means that the Universe is not simply accessible, but that its behavior should be ascertainable using the laws and theories governing terrestrial and solar phenomena. Studies of the Universe therefore belong to the intellectual enterprise—to the universe of ideas. Cosmology is possibly the grandest member of that enterprise, for, as I noted earlier, what could be grander than attempting to understand—to demystify— the Universe?

And, by the way, Kirchhoff's discovery consigned to the intellectual trash bin the 1842 claim by the French philosopher August Comte that the chemical composition of stars would never be known. It is also the foundation for Albert Einstein's much later remark that the most incomprehensible thing about the Universe is that it is comprehensible.

Many elements were discovered by means of spectroscopic analysis. Working with his German friend Robert Bunsen, inventor of the Bunsen burner, Kirchhoff discovered the previously unknown elements cesium and rubidium. Helium, whose nucleus plays a vital role in the generation of energy in stars like the sun (which I will discuss in the next chapter), was unknown prior to 1878. Its existence was inferred that year from the analysis of absorption lines in the sun's spectrum by the English astronomer Joseph Lockyear, but it was not until 1895 that the Scottish chemist William Ramsay detected it on the earth.

The work of the 19th century spectroscopists discovering and identifying what are now known as chemical elements continued well into the 20th century. A total of 83 *stable* elements are now known, the lightest being hydrogen and the heaviest bismuth. There are also many heavier ones, all *unstable*, that radioactively decay into lighter elements by the emission of various particles.

An element consists of a single *atom*, which in turn is composed of *neutrons* and an equal number of *electrons* and *protons*. Neutrons, particles with no electric charge, are slightly heavier than protons, each of which has one unit of *positive charge*. Electrons, on the other hand, are far less massive than protons and have one unit of *negative charge*. Atoms are electrically neutral because they contain equal numbers of protons and electrons (the sum of all the proton charges is equal and opposite to the sum of all the electron charges, so that their total sum is zero, ensuring zero charge and thus electrical neutrality). Two examples are hydrogen, which has one proton, one electron, and no neutrons, and bismuth, which contains 83 protons and electrons and 126 neutrons. The chemical properties of atoms are completely determined by the number of electrons they contain, independent of the number of neutrons and protons.

Molecules consist of two or more neutral atoms. They are held together by the same electrical forces that bind electrons to the protons in an atom's nucleus; these are also the forces that determine the maximum height of mountains and prevent a chair from collapsing when a person sits on it. In contrast with neutral atoms and molecules, *ions* are atoms or molecules to which one or more electrons have been added or subtracted, so that they are, respectively, negatively or positively charged.

In terms of these fundamental constituents, the previous statement about the composition of the Universe may be rephrased as follows: despite the best efforts of science fiction authors, the only atoms, molecules, and ions so far detected anywhere else in the Universe are those encountered on the earth and in the solar system. That this should be so follows directly from the Big Bang theory, as I describe in the penultimate chapter of the book. The atmospheres of stars contain mostly atoms and ions, whereas molecules tend to be more concentrated in interstellar space. All can emit and absorb radiation, and the spectra for enormous numbers of atoms, ions, and molecules have been determined and cataloged.

Why (and how) these microscopic objects emit and absorb radiation only at particular wavelengths is a consequence of the quantum nature of their "structure," a topic I consider in the last section of this chapter. It suffices here to note that the popular picture of an atom as a miniature solar system, wherein the protons and neutrons form a tiny nucleus with the electrons moving in certain orbits around it, is incorrect: electrons do not move in "orbits" of any kind, well-defined or not.

Blackbody Radiation and Photons

To almost all of its practitioners, late 19th century physics was a seamless construct, with only a few unimportant "loose ends" to be tidied up. No one living then could have conceived of the revolutions in theory that were about to occur. The paradigm changes grew out of two sets of measurements.

The first was related to the wave nature of radiation as specified by Maxwell's equations. Every physicist knew that a wave was a disturbance in a medium. But it was also known that most of the solar system was empty of matter, which led to the question, "What is doing the waving when light travels from the sun to the earth?" The answer was a hypothesized medium, both transparent and motionless, variously called the aether, or the ether, or the luminiferous ether. Postulated to exist throughout space, the earth was believed to be swimming in it. Its presence would have been manifested as an "aether wind," whose effect would have been changes in an object's speed, depending on whether the object's motion was along or opposite to the earth's as it orbited the sun.

Sunlight is one object whose speed should have exhibited a change, and in 1881 the American scientist Albert Michelson carried out an experiment designed to detect the effects of the aether wind on light.

His result was no change in speed; his conclusion was that the aether did not exist. Unfortunately for him, the experiment was described in a paper published in a little-read journal, especially by the then giants of science, and when their attention was directed to it, they cither ignored it or queried the validity of the result. This response was sufficient motivation for Michaelson to repeat the experiment. He did so with a collaborator, the American physicist Edward Morley. Using improved equipment, they found a null result again. They published their results in 1887, and this time the giants of the time did pay attention. One, Lord Kelvin (William Thomson), noted in a 1904 lecture that the null result was the only objection to an otherwise exemplary theory of radiation. Little did he know what Albert Einstein was to propose just 1 year later.

The second set of experiments that led to paradigm change were measurements of *blackbody radiation*. A blackbody is a perfect absorber of radiation; it is well approximated by certain types of ovens. Suppose that radiation exists in a cavity inside a blackbody and that the system of radiation and blackbody is in *equilibrium* at a temperature T (the units in which T is measured are *Kelvin*, denoted K; they are defined in the next chapter). Equilibrium means that the walls of the blackbody surrounding the cavity emit as much radiation as they absorb; it is the condition necessary for the radiation to be blackbody.

In principle, all wavelengths are present in blackbody radiation, and the relevant experiments consisted of measuring, at different wavelengths and temperatures, the intensity of the radiation emanating from a small hole in the blackbody. The decisive measurements were made in 1900 by two teams of German experimenters, first by Otto Lummer and Ernst Pringsheim, and a little later by Heinrich Rubens and Ferdinand Kurlbaum. As this was the heyday of Maxwell's theory and of thermodynamics (the science of heat), it was expected that the experimental results would be easily explained, with the data being fitted by a nice theoretical curve derived from these theories. This expectation was unfulfilled: none of the attempts to explain the results using the extant theories was successful, two even predicting infinite intensities (at very long and at very short wavelengths).

These infinities were scientific catastrophes, caused, though no one yet knew it, by theorists assuming, as they had always done, that the energy of the radiation in the cavity was continuously distributed over the wavelengths, just as occurred in the sun's spectrum.

Into this catastrophic situation stepped the German theorist Max Planck, who fitted the data by means of an *ad hoc* mathematical formula. He derived this formula from the standard framework by making a "crazy" assumption, one to which he could never reconcile himself, even after he had won the Nobel Prize for it! His assumption was that the radiation in the blackbody cavity could be emitted or obsorbed only in certain *discrete* amounts or bundles, which he called *quanta*. In particular, for each frequency, he proposed that the energy was proportional to the product of the frequency f and an integer (or equivalently, that it was proportional to an integer divided by the wavelength λ: recall the formula $c = f \times \lambda$). By fitting his *ad hoc* curve to the data, he extracted the value of the constant of proportionality occurring in the energy/frequency expression; it is known as Planck's constant, and the radiation energies that are proportional to it are said to be *quantized*. Each discrete bundle of blackbody radiation is therefore a *quantum of energy*. And, as I explain below, quantization of energy is not limited to blackbody radiation: it is a general characteristic of all microscopic systems.

Despite its success in fitting the data, Planck's quantum explanation was not well received: its violation of the dearly held tenet that the energy of radiation had to be continuously distributed over wavelengths was too much for the scientific community to accept—and, as noted, Planck himself was never comfortable with it. However, all other attempts to explain the data failed, surely a sign that new thinking was required.

The only other person able to think outside the continuity-of-radiation box was Albert Einstein. In 1905, 5 years after Planck's quantum hypothesis, Einstein extended it to the phenomenon called the photoelectric effect, in which radiation impinging on a metal could cause electrons to be emitted from its surface. Maxwell's radiation theory could not account for the data, whereas Einstein did so by postulating that in this circumstance, just as in blackbody radiation, the impinging radiation behaved *as if* it were particulate, composed of bundles or quanta of energy. That electromagnetic radiation actually did behave this way was confirmed experimentally in 1923 by the American physicist Arthur Compton. Radiation quanta were subsequently renamed *photons* by the American physical chemist Gilbert Lewis, who used it in the title of a 1926 scientific paper. The name caught on immediately and has become part of the scientific nomenclature.[7]

So, after nearly 300 years, the way scientists thought about the nature of light had come full circle, back to the corpuscular quality accorded it by Newton. No further changes in this regard have occurred, nor are they likely to do so: it is firmly believed that all electromagnetic radiation consists of photons. There is only one problem: if sunlight is composed of photons, how are its wave-like properties to be understood? A wave is a wave, a particle is a particle, and presumably a particle will never wave.

Particles do not require a medium in which to propagate, but Young's interference experiment requires waves. Quandary! This seeming paradox is explained by the fact that light is really a quantum, and not a Maxwellian, phenomenon.

The new paradigm—the "truth" if you like—is that depending on how it is observed, light (and all radiation) can be a particle *or* a wave; which aspect will be manifested depends on the experimental circumstances. All electromagnetic radiation is indeed made up of photons, but they can behave at times as if they are waves, not particles. To understand how this can occur requires delving further into quantum theory than I will undertake in this book. If you wish to learn more about the quantum nature of light and radiation, I suggest trying the semipopular account presented in Richard Feynman's beautiful little monograph titled *QED* [Feynman (1985)].[8]

Although my next topic is the quantum nature of atoms, molecules, and nuclei, I don't want to go there without emphasizing that blackbody radiation is much more than a convenient vehicle for introducing the concepts of quanta and the discreteness of energies. For instance, the radiation emitted by our own star, the sun, is approximately of blackbody character, as you will see in Figure 11 of the next chapter.

More significantly, the cosmic microwave background radiation is so perfectly blackbody that the deviations from it are at most 1 part in 10^5. The presence of the CMB—mainly in the microwave and infrared portions of the spectrum—is the single most important datum among several that support the Big Bang origin of the Universe. The tiny deviations in the CMB from a pure blackbody spectrum are highly important in their own right, shedding light on galaxy formation and confirming the recent conclusion that the Universe is not simply expanding, it is also accelerating. I devote much of Chapter 7 to the origin and analysis of the CMB.

Quantum Concepts

Starting with Planck's *ad hoc* quantizing of blackbody radiation in 1900, it took a little more than a quarter-century to create a complete quantum theory of microscopic phenomena, including a quantized theory of electromagnetic radiation. An intermediate step in this evolution was another *ad hoc* creation, the 1913 model of the hydrogen atom of the Danish physicist Niels Bohr that featured the now-abandoned concept of electron orbits. Such orbits still appear in places, for example, as the oblique circles seen on the yellow signs warning of the presence of radioactivity. Bohr's orbits function only as pictorial aids: they do not occur in modern theories.

By 1927, the major components of quantum theory were in place, created by a group of brilliant theorists, among them Max Born, Wolfgang Pauli, Pascual Jordan, Paul Dirac, Erwin Schroedinger, and Werner Heisenberg (of *uncertainty principle* fame, a principle seemingly misunderstood by some writers who produce "popularizations" often accepted by gullible seekers of wisdom). Many experimental results, some new, some old, were explained by the new theory, which has replaced Newtonian mechanics and Maxwellian electrodynamics as the once-reigning paradigms. Quantum theory accounts for the structure of atoms, molecules, and nuclei, as well as for the properties of solids, electrical conductivity, devices of all sorts, and a host of other applications. It is THE framework describing matter on the microscopic scale.

There are several features of quantum theory I wish to emphasize here. Of the least significance to the journey undertaken in this book, but one that will help explain the absence of electron orbits, is the role of probability. Orbits are a nonfeature because the best that quantum theory can do in this regard is to provide the *probability* that an electron (in an atom or molecule, say) will be at a particular location: depending on the system and its internal energy, some locations are more—even much more—probable than others, but none are certain.

One way to understand the inability to predict the exact location of an electron is by considering the following experiment. Suppose you were to try measuring its position in an atom by using radiation to locate it. Unfortunately, the radiation would bring so much energy into the system that the electron would be knocked out of the atom. Indeed, no experiment attempting to measure an electron's location in an atom will leave it intact. So, if experiments cannot measure an electron's position, then theory should not be able to predict its location either, and it doesn't.

Probability is inherent in some aspects of quantum theory, but not in all. Internal energies can be calculated and measured to arbitrary accuracy, as can electromagnetic properties, which are intimately related to the structure of microscopic systems. In ordinary speech, "structure" implies both a building of some kind and a form or shape. In microscopic physics or chemistry, however, structure refers in part to the number and type of constituents that make up an object (e.g., the number of atoms in a molecule; how many electrons, protons, and neutrons are in an atom; etc.). "Structure" also refers to the internal energies of a microscopic system and to entities known as the *states* of the system. Apart from nomenclature, the latter play will play little role in my discussion, but the former are crucial to it.

As I noted before, the internal energies of microscopic systems occur only in discrete, quantized amounts. Such a statement may not seem momentous. Yet, from the time when energy quantization was first accepted as a fundamental attribute of quantum systems up to the present day, the discreteness of quantum energy levels has been an extraordinary quality. One reason is that discreteness contrasts sharply with the behavior of energy in the macroscopic systems that form our visual environment. In the realm of human experience, energy always takes on a range of *continuous* values. Two examples are *kinetic energy*, which varies as the square of the speed, and *gravitational energy* on the surface of the earth, which varies linearly with height. The continuous range of each of these two energies follows from the fact both speed and height can each change continuously: neither is restricted in value.

The discrete internal energies of microscopic systems are referred to as *energy levels*, the totality of which forms an *energy spectrum*, in analogy to the radiation spectrum of Figure 8. Visible light is broadly decomposed into the six colors from red to violet, an ordering from longer to shorter wavelengths. An analogous ordering is usually imposed on energy-level spectra, except that they always run from the minimum value to the largest. These spectra are normally displayed via *energy-level diagrams*, wherein the levels are indicated by short horizontal lines. The line depicting the level with the least energy is at the bottom of the diagram, and the others are shown sequentially above it, ending with the level having the greatest energy represented by the line at the top.[9]

Figure 9 is such a diagram for a system with five energy levels (a real system could have more or fewer levels: this is just an illustrative example). The bottom line in the figure represents the *ground-state* energy: there cannot be a level in any quantum system having an energy lower than that of its ground state. Each of the lines above the lowest signifies an *excited-state* energy; they are denoted the first, second, third, etc, excited-state energies. This labeling extends to the subscripted letters E, which indicate the values of the energy, with the subscript "0" identifying the ground-state energy.

Every microscopic body tends to be in its ground state, whose energy E_0 is specific to the body. A microscopic system can be put into one of its excited states by absorbing energy. This can occur by collisions with other microscopic bodies or by ingesting a photon ("ingesting" is an appropriate description, as the photon transfers all its energy to the system and disappears). Once in an excited state, the system eventually will make a *transition* or *decay* to either the ground state or a lower-lying excited state.

When the transition to, or the decay of, an excited state occurs via electromagnetic radiation, a photon is either absorbed or emitted by the system. Such absorption or emission of photons is the origin of dark-line and bright-line spectra, respectively, with

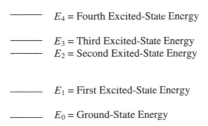

Figure 9. An energy-level diagram showing the five levels of a hypothetical microscopic system. The energies are denoted E_0, E_1, ... E_4, with the ground-state energy E_0 being the lowest (and numerically largest in magnitude), while the positions of the horizontal lines are a symbolic representation of their values.

the wavelengths being determined by the energy differences between the initial and final states of the microscopic system. In particular, the energy of such a photon is equal to the latter energy difference. Just as in the case of blackbody radiation, photon energies are inversely proportional to their wavelengths. Hence, the wavelength λ of an emitted or absorbed photon is inversely proportional to the difference in energies between the higher and lower states that are linked by the photon: $\lambda \propto 1/(E_{\text{higher}} - E_{\text{lower}})$.

These comments are illustrated in Figure 10, wherein the energies of the system are represented by the horizontal lines and the photons are indicated by the vertical dashed arrows. An upward arrow specifies absorption (energy is added to the system, thereby raising it to a higher level), whereas a downward arrow means emission of radiation, with a corresponding shift to a lower level.

Figure 10a depicts the general situation for the case of electromagnetic transitions taking place between just two levels, E_n and E_m. If the system is in level E_m, it makes a transition to higher level E_n by absorbing a photon (upward arrow), whereas if the system starts in the upper level, it decays to the lower one by emitting a photon (downward arrow). This part of the figure demon-

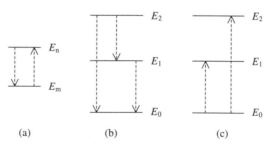

(a)   (b)   (c)

Figure 10. Energy-level diagram representation of the emission and absorption of electromagnetic radiation, whose photons are indicated by the vertical dashed lines with an arrow. (a) An upward arrow specifies *absorption*, with corresponding *excitation* of the system from the lower level E_m to the higher level E_n, whereas a downward arrow indicates *emission*, with a *decay* from the higher level to the lower one. (b) The second excited state is graphically seen to be able to decay either to the first excited state or to the ground state, whereas the first excited state can decay only to the ground state. (c) Transitions out of the ground state via absorption of photons can go to either of the two excited states.

strates why the wavelengths for emission and for absorption between the *same* levels are equal (recall Fraunhofer's discovery that the wavelengths of the dark and bright lines of sodium were the same).

Figures 10b and 10c enlarge the arena by specifying the three lowest-lying levels of a system: the ground-state and the first two excited-state energy levels. Both absorption and emission transitions are shown. Normally only one photon at a time is involved, so the decay from level E_2 can go either directly to the ground-state level via single photon emission or as a *cascade* of two sequentially emitted photons, the first to level E_1, the next from E_1 to E_0.

It is possible for the energy difference between a pair of levels in one system to be equal to that in another system, but the collection of differences—in effect the whole spectrum—is unique for each atom or molecule or nucleus. This is generally true for segments of the spectrum as well. Hence, a quantum system's unique signature is obtained by measuring the spectrum of its emitted photons, even—and especially—when the complexity of systems composed of many smaller bodies makes it impossible to calculate energies with sufficient accuracy. It is this unique signature, explained by quantum theory, that helped open the way toward understanding much about stars and the Universe in general.

4. Stars: Attributes, Energetics, End Stages

"Twinkle, twinkle, little star/How I wonder what you are. . . ." For some, these lines could evoke childhood memories, but to those with an interest in astronomy, they may suggest a truth, a miniaturization, and a major scientific achievement. Stars do "twinkle," but only because they are observed through the earth's turbulent atmosphere. Like the random movements of particles suspended in liquids, the twinkling is a result of the motion of atoms and molecules. This turbulence affects the seeing from earth-based optical telescopes, but through the use of "adaptive" optics, telescope mirrors can be made flexible enough to partially compensate for the atmospheric distortion. "Little stars" appear to be miniaturized, but only because of their enormous distance from the earth: apart from white dwarfs, neutron stars, and nonmassive black holes, stars are huge. An example is the sun, whose mass and volume, respectively, are roughly 300,000 and a million times that of the earth.

The scientific achievement concerns the "How I wonder" phrase: the lack of knowledge it represents is no longer a deficiency. From the analyses of Gustav Kirchhoff in the mid-19th century to the very recent efforts of many physicists, astronomers, and astrophysicists, the detailed properties of the sun and, by implication, stars in general have become well understood. Not surprisingly, our system's star has been the test case in the attempts to pin down details of stellar formation, overall composition, energy generation, luminosity, temperature, and evolutionary tracks and end stages. The sun is just one star in a gigantic population of them, and much more has been learned by examining aggregations of stars. A tool for doing this has been the Hertzsprung–Russell diagram, which you will encounter later in this chapter. But first, on to stars!

A star is a giant, luminous sphere of matter and radiation. Most stars are in hydrostatic equilibrium, which means that the attractive force of gravity that would cause the matter to fall inward is balanced by the outward pressure exerted by its contents. A failure to maintain equilibrium can lead to the occurrence of violent events such as supernovas. Fortunately for us, the sun has been in equilibrium for roughly 5 billion years and will continue that way for at least the same length of time, until it becomes a red giant.[1] As I will demonstrate later in this chapter, the 10-billion-year estimate for the sun's lifetime follows from knowledge of its luminosity, the number of protons it contains, and the energy released by the fusing of protons to make the nuclei of helium atoms.

A star's *luminosity* is the amount of energy it radiates per second, but how bright it appears depends on its distance away. The larger that distance, the greater the decrease in brightness: the pole star Polaris is roughly 13 pc away, yet it is no brighter than a lighted candle at a distance of 1.6 km. You can understand qualitatively how this happens by thinking in terms of photons. The greater the number of visual photons that enter your eye, the more intense—or bright—a source of radiation will be. Imagine a closely knit group of photons being emitted from a small surface area of the source. Their lack of perfect collimation means that they will gradually move away from one another as they travel, and if they travel very far, only a small percentage of them will reach you. Fewer photons means lower intensity, which means decreased brightness compared with what it would be were you close to the source.

A more quantitative description uses the fact that brightness decreases as the square of the distance away. A 100-W bulb at a distance of 100 m is as bright as a 400-W bulb 200 m from you, even though the luminosity of the second one is four times greater than the first. Astronomers and cosmologists measure the *apparent luminosity*, ℓ, but they need to know the value of the actual luminosity L. In view of the falloff just noted, the relation between them is $\ell \propto L/D^2$, or equivalently, $L \propto \ell \times D^2$, where D is the distance to the source of radiation (and the constant of proportionality is known). L can be calculated once ℓ and D have each been determined, and this has been a motive in the search for methods to measure stellar distances accurately.

The unit for both stellar and galactic luminosities is the sun's luminosity L_{Sun}, and the successful attempt to determine it is the prime application of the preceding formula. Only a measurement of the apparent luminosity is needed, since D_{ES}, the distance to the sun, is already known. To eliminate atmospheric absorption, ℓ_{Sun} was measured just above the earth's atmosphere. The energy falling each second on a unit area at that height is known as the *solar constant*; its value is 0.137 W per square centimeter, or roughly one-seventh of a watt/cm^2, much weaker than the energy radiated per second by household light bulbs.

The calculation of L_{Sun} from the solar constant involves multiplication by the square of the earth–sun distance. Performing the requisite calculation leads to $L_{Sun} = 3.85 \times 10^{26}$ W, essentially the number quoted in the discussion of powers of ten in Appendix A. How potent is this amount of energy? Abell (1975), among others, notes that if a giant bridge of ice, 3 km by 1.5 km, were constructed between the earth and the sun and then all of the sun's luminosity were directed along it, the ice would melt in 1 second!

The Kelvin Temperature Scale

Luminosity is one of the attributes characterizing the outer layer of a star—its *photosphere*. Others are color, contents, and temperature. Like luminosity and contents, surface temperature is an attribute astronomers try to measure. Stellar temperatures are expressed in degrees Kelvin (K), rather than the more familiar ones of degrees Fahrenheit (°F) or degrees Celsius (°C). Fahrenheit is commonly used in the United States, Celsius almost everywhere else. The scales of these two temperatures are defined through the freezing and boiling points of water, which under normal conditions freezes at 0 °C or 32 °F and boils at 100 °C or 212 °F. Degrees expressed in one of these scales are related to those in the other by a simple mathematical formula,[2] which for temperatures somewhat above the boiling point of water leads to the result that °F $\cong 2 \times$ °C.

Temperature, which you are probably used to reading from a thermometer, is actually a measure of the average speed of the molecules forming the substance whose temperature is being taken.[3] Heat the substance, and the molecules speed up; cool it, and they slow down.

This latter feature plays an essential role in the definition of the Kelvin temperature scale. Zero on this scale is defined as the temperature at which all motion ceases. It is written as 0 K, where the degree symbol ° is deliberately omitted. 0 K is known as *absolute zero*, and temperatures expressed in the Kelvin scale are often referred to as *absolute* temperatures. In terms of the more familiar units, absolute zero occurs at −273.16 °C or −459.69 °F, temperatures far colder than any encountered during the coldest Arctic winters. Scientists use the Kelvin scale because it enjoys a special advantage over the other two: unlike them, the reading does not depend on the material in a thermometer.

An approximate value of the temperature in degrees Kelvin is obtained from the familiar ones by adding 273 degrees to the Celsius reading and 460 degrees to the Fahrenheit value. Thus a temperature of 25 °C (77 °F) is nearly 300 K. As long as the temperature is very high, the C and K values are about the same, since in this instance the approximately 275-degree difference between them can be ignored. In contrast, temperature values expressed in °F and in K will differ by roughly a factor of two, as temperatures expressed in °F are almost twice as large as in °C.

Measuring Surface Temperatures: The Blackbody Method

There are various indirect methods by which photosphere temperatures are measured—or at least estimated—and I shall describe two of them, one of which uses blackbody radiation, the other photosphere spectra.

In the blackbody method, the star is assumed to be an emitter of blackbody radiation. Its luminosity is then given by a blackbody formula, wherein L is proportional to the square of its radius R and the fourth power of its surface temperature T_{surf}: $L = C \times R^2 \times T_{surf}^4$, where C is the known constant of proportionality. In general, stellar radii are not known, but if L and T_{surf} have been determined independently, this relation can be used to evaluate R. The sun, on the other hand, is a star for which L and R are known, so its surface temperature is obtained from a simple calculation. The result is $T_{surf} = 5780\,K$, the currently accepted value. By terrestrial standards, it is a very high temperature (hot coals are about 1250 K, an incandescent light bulb filament is roughly 2000 K), but not by stellar standards, as you will see later in the Hertzsprung–Russell diagram.

Because the value of T_{surf} is derived from the blackbody assumption, it is important to test its validity: does the sun behave like a blackbody emitter, or is the assumption one of unjustifiable convenience? A direct answer to this question can be obtained by measuring the sun's spectrum and comparing it with that of a blackbody at the temperature T_{surf}. Such measurements have been made; a comparison is shown in Figure 11.[4] In it, the sun's spectrum has been smoothed to eliminate the dark-line dips arising from atmospheric absorption.[5] While far from perfect, the agree-

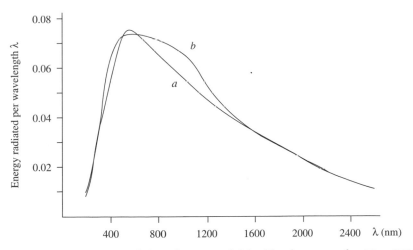

Figure 11. Comparison of the theoretical blackbody curve for $T = 5780\,K$ with the sun's spectrum, with the wavelength λ, in units of nanometers, specified on the horizontal axis, and the amount of energy radiated per wavelength indicated on the vertical axis (the energy units, being unimportant for our purposes, are unspecified, though the relative amounts are correct.) Curve a: the theoretical blackbody spectrum; curve b: the sun's spectrum (adapted from the Solar Spectrum Web site).

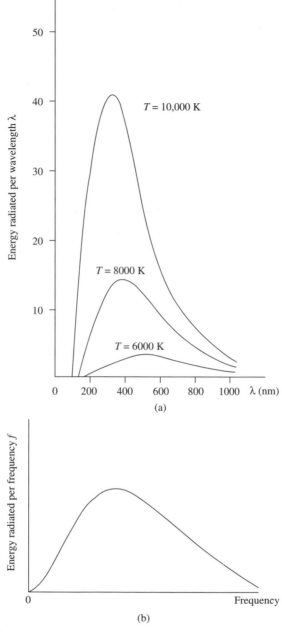

Figure 12. (a) Comparison of the energy per wavelength of blackbody radiation for $T = 6000\,\text{K}$, $T = 8000\,\text{K}$, and $T = 10,000\,\text{K}$, plotted against wavelength λ. (b) Generic blackbody curve showing the variation with frequency f.

ment between the two curves is quite good overall, implying that the sun's surface *is* reasonably well described as a blackbody emitter.

Figure 11 displays several noteworthy features. First, the maximum in both the data and the theoretical curve occurs at approximately 500 nm, which is a wavelength in the green, as previously stated.[6] Second, the sun's smoothed intensity shows an increase relative to the blackbody curve as one goes further into the ultraviolet (UV) and infrared (IR) wavelengths, a result to be contrasted with some decreases at those wavelengths—due to atmospheric absorption—when the spectrum is not smoothed.[7] Third, the theoretical curve (and also the sun's spectrum) approaches zero for large as well as small wavelengths, in accord with the claim that Planck's blackbody radiation formula is free of infinities.

In the general case where neither a star's radius nor its surface temperature is known, but enough of its spectrum has been measured, the preceding comparison can yield an estimate for T_{surf}. (Values determined this way are deemed *effective* temperatures.) The procedure is to match a series of blackbody curves to the measured (and possibly smoothed) spectrum until a best fit has been obtained. Suppose now that you have been asked to extract a value of T_{Surf} in this way, and that you have both the spectrum and the requisite computer programs. You would quickly learn that blackbody curves are sensitive to the value of T used to generate them, as shown in the wavelength version of Figure 12 (for comparison, a generic example of the frequency version is also displayed in the figure). In precomputer days, such an analysis was tedious to carry out; nowadays, sophisticated software relieves the tedium.

Measuring Stellar Temperatures: Spectrum Method

The second method for estimating photosphere temperatures involves stellar spectra. Photospheres typically contain photons, atoms, molecules, and ions (plus electrons, which are not germane to the analysis). By absorbing photons of the relevant wavelengths, the atoms, molecules, or ions can make transitions from their ground states into excited levels and from one excited state to another. These excitations produce the dark lines in the spectrum of a star. The *relative intensities* of the dark lines, which are the signatures of the transitions, are a measure of the *relative populations* in the excited states. There are theoretical formulas that relate the relative populations to the photosphere's temperature, so once the relative intensities of the dark lines are measured, values of T_{surf} can be and have been deduced.

For this method to be practical, T_{surf} must be large enough to ensure detection of the relevant levels. Why does the value of T_{surf} matter? Here's the argument. Excited levels in atoms, molecules, and ions are generally populated by absorbing photons of higher rather than lower energy, and as the value of T_{surf} increases, so do the number of higher-energy photons. Absorption of more photons creates greater numbers of excited levels and, correspondingly, greater intensity of the dark lines. To help explain the relation between T_{surf} and photon number, I will use blackbody radiation as an analogy.

For blackbody radiation, the number of photons (per unit volume) is proportional to T^3, so their number increases rapidly with T. Also, the number of higher-energy photons—which have the smaller wavelengths—increases more quickly with T than does the number of lower-energy ones. Reference to Figure 12a should make this clear. So, in the case of blackbody radiation, having enough photons with energies to excite a detectable number of excited states requires a sufficiently high temperature. The same concepts apply to stellar photospheres even if their photons are not blackbody: as T_{surf} increases, so do the number of high-energy photons, and so will the probability of detecting the dark lines in the star's spectrum.

Stellar temperatures determined in the preceding manner are known as *excitation* temperatures. Knowledge of stellar spectra is the essential ingredient in this method, and much of 19th century astronomy was devoted to measuring them. Spectra fell into distinct groups, leading to a taxonomy of stars expressed in terms of their *spectral class*. Spectral class is a shorthand for identification via the most abundant atoms, molecules, and ions in the photosphere; it is used in constructing the Hertzsprung–Russell diagram.

Spectral classes are designated by the letters O, B, A, F, G, K, and M. (A mnemonic for the spectral class letters is "Oh Be A Fine Girl/Guy, Kiss Me.") Class O contains the hottest (blue) stars, with temperatures as high as 40,000 K, while the coolest (red) stars are members of class M, whose lowest temperatures are about 2000 K. Each class has ten subdivisions. The sun, whose spectrum shows the presence of ionized calcium plus both neutral and ionized metals among other substances, is a member of spectral class G. (Readers of older, space-opera type of science fiction may recall that the sun was occasionally referred to as a class G2 star.)

Stellar Energy Sources: Background

Up to now, my emphasis has been on external properties of stars such as surface temperature, radius, luminosity, spectra, and photospheric composition. They are measurable attributes, at least in principle (although accurate determination of luminosity is far from straightforward and will be considered later). Internal properties, on the other hand, are not measurable, yet determining them is crucial to gaining an understanding of stars. Among the features astrophysicists would like to know are the types of particles present in different regions of the interior; the temperature, pressure, and density distribution throughout the stellar volume; the mechanisms by which energy is generated and eventually brought to the surface; and how stellar mass influences these answers. Knowledge of these features and others has been gained by constructing stellar models based on terrestrial laws and theories. The laboratory in which such models have been tested is the sun.

The first scientific attempt to account for the sun's radiant energy—and by inference that of other stars—was based on a mid-19th century proposal of the German scientist Heinrich

Helmholtz. He assumed that the sun was slowly contracting under the influence of its own gravity. Gravitational theory states that a consequence of the resulting decrease in volume is an increase in internal energy, an increase that Helmholtz proposed as the source of the energy lost through radiation from the sun's surface. Eventually, of course, the process would end with collapse of the sun. Numbers were put into this solar model in the early 1860s by William Thomson (who later became Lord Kelvin, for whom the temperature scale is named). Using the mass of the sun and its estimated luminosity, Thomson deduced a solar lifetime of about 20 million years.

However, even then it was realized that 20 million years was incompatible with the much longer time spans needed for the occurrence of geological phenomena (and biological evolution as postulated in Darwin's theory): the sun was not supposed to be younger than the earth! Radioactive dating methods have shown in the 20th century that the earth, moon, and meteorites have ages of approximately 4.6 billion years, a number believed to be close to the age of the solar system. The sun must be at least this old as well.

For the sun to have lived about 5 billion years or so while continuously radiating photons from its surface, its internal pressure must be sufficiently high to have prevented gravitational collapse. (Were the internal pressure zero, then at the sun's inferred central density of approximately $1.5 \times 10^5 \, kg/m^3$, collapse would occur in about 30 minutes![8]) Because pressure and temperature tend to be linearly related, the large internal pressure implies that the internal temperature of the sun must also be large. A large internal temperature means that the particles in the sun's core are moving at high speed, high enough that they can collide with one another. Such collisions, or *reactions*, are expected to be the energy sources generating the photons that power the observed luminosity.

Photons are produced terrestrially by chemical reactions and nuclear reactions. Photon-producing chemical reactions occur in familiar settings: the striking of a match, the burning of wood or paper, the explosions of fuel in the cylinders of automobiles. Chemical reactions, however, do not supply sufficient energy: they are too weak by roughly a factor of a million. Nuclear reactions, familiar from nuclear weapons and nuclear power plants, not only generate enough energy to power the sun and stars, but they also have been universally accepted as the photon-production mechanism. Entities other than photons are produced as well, as you will see shortly.

The first experimental studies of nuclear reactions were carried out in the 1930s. In 1932, the Englishman James Chadwick discovered the neutron and the American Harold Urey discovered the deuterium atom (its nucleus has a neutron in addition to the lone proton). These discoveries were crucial to the 1939 analysis of the German-American physicist Hans Bethe that identified the reactions that lead to stellar energy production. In particular, he established, in Nobel Prize–winning research, the nuclear reactions that are relevant for stars in hydrostatic equilibrium. They involve the "burning" of four protons to make an alpha particle, the nucleus of the helium atom.

Proton burning occurs along one set of paths in stars whose masses are comparable with that of the sun, whereas for more massive stars it occurs along another set in which a carbon nucleus acts as a catalyst. Some of the details of this process and its consequences for solar-mass stars are the subject of the next section; they are also relevant to later chapters of the book. Readers who wish to learn more than I shall describe below might consult the books by Clayton, Harrison, Phillips, or Webb listed in the Bibliography.

Stellar Energy Sources: Some Details

There are two kinds of attractive forces that stabilize atoms. Electrical forces bind the electrons to the protons in the nucleus, although instead of occupying orbits, the electrons may be thought of as forming a "cloud" surrounding the nucleus. The neutrons and protons in the nucleus are bound together by attractive forces strong enough to overcome the electrical repulsion that would otherwise pop the protons out of the atom. This attractive nuclear force (also known as the *strong* force) stabilizes nuclei in a second way: it usually prevents neutrons from *decaying* into other particles via a process known as *beta decay*. Beta decay is mediated by the *weak* force, which causes unbound neutrons to transform or decay into a proton, an electron, and an *antineutrino*. Decays and antineutrinos are essential elements in the story I relate in this and subsequent chapters.

In physics, decay refers to the transforming of an unstable body into two or more bodies, at least one of which is stable. If, among the resulting set of two or more bodies, there is again an unstable one, it will also decay, possibly producing another unstable body, and so on, until there is a final set of stable bodies. The transforming event often *creates* the new matter from the old, as in the case of beta decay: the neutron does not contain the new bodies in it, as if they were just waiting to get out of jail.

Each unstable body is characterized by a measurable *lifetime*, which is the time it exists before decaying. Quantum theory yields formulas for lifetimes, and for simple enough systems the lifetime values calculated from these formulas are in good agreement with experiment (the accuracy tends to decrease as the complexity of the decaying system increases). The hydrogen atom (a proton and an electron) is an example of a simple system, and the calculated lifetime of its first excited state against photon decay to the ground state—approximately 1.6×10^{-9} sec—is in excellent agreement with the measured value.

A collection or sample of identical unstable bodies is characterized by a different intrinsic time known as its *half-life*. It is the time at which half of the sample will have decayed. After one half-life, the remainder of an original sample of radioactive nuclei will continue to decay and will therefore be dangerous, unless the sample is shielded, for example by lead.[9]

The neutron has a lifetime in free space of approximately 15 minutes; a collection of them has a half-life of roughly 10 minutes. Transformation into a proton, an electron, and an antineutrino is deemed beta decay because at one time an electron was referred to as a beta particle.[10] Physicists denote decay processes using arrows, in analogy to chemical reactions: the symbol for the decaying particle appears to the left of the arrow, on whose right side are the symbols for the decay products. Neutron decay is thus written as

$$n \rightarrow p + e^- + \bar{v},$$

where n stands for neutron, the arrow $\rightarrow$ symbolizes both the decay process and the direction in which it proceeds, p means proton, e⁻ indicates an electron (the superscript minus sign denotes a negative charge), and $\bar{\nu}$ denotes an antineutrino (ν is the lower-case Greek letter nu).

The antineutrino, which plays an important role in stellar nuclear reactions, is an *antiparticle*, here the antiparticle to the *neutrino*, whose symbol is ν. Antiparticles generally differ from their corresponding particles only in the sign of their electrical properties; the two masses are the same. The antiparticle to the electron is the *positron*, denoted e⁺; it has a positive charge equal to that of the proton, whose own antiparticle is the negatively charged *antiproton*, denoted $\bar{p}$. Particle–antiparticle pairs can be created in energetic collisions by turning some of the energy into mass; on contact, the members of the pairs annihilate one another, producing photons. Some uncharged, massless particles are their own antiparticles—the photon is an example—whereas others such as neutrinos have distinct antiparticles. The theoretical prediction that ν and $\bar{\nu}$ are distinguishable has been verified experimentally. An important theoretical/experimental fact is that electrons are always paired with antineutrinos, while positrons are paired only with neutrinos.

You may be wondering at this point why some particles decay and others do not. Decays can only occur if the unstable particle has a mass greater than the sum of the masses of the decay products: in the language of chemistry, the decay process is *exothermic*; that is, one in which energy is released. (Exothermic *nuclear reactions* are the only ones that occur in the buildup of intermediate-mass nuclei in stellar interiors.)

The amount of energy that can be shared among the decay products is given by a variant of Einstein's $E = Mc^2$ formula. It is the product of c^2 and the *difference* between the mass of the unstable particle and the sum of the masses of the decay products. For beta decay, the expression for the available energy is $[M_n - (M_p + M_e + M_\nu)] \times c^2$, where the subscript on each mass symbol M stands for a particle taking part in the decay (recall that particles and antiparticles have the same mass). Were the sum of the masses of the proton, the electron, and the antineutrino greater than the neutron mass, the factor in the square brackets [ ] would be negative, the process would not be exothermic, and the decay would be forbidden. It is this type of argument that forbids isolated protons from decaying into neutrons: because the proton mass is *less* than the neutron mass, the factor in the analogous square brackets is negative. Nonetheless, the transformation of a proton into a neutron (an analogue of proton beta decay) does occur in one of the proton-burning reactions described below.

The neutrino mass is one of those present in the square brackets for neutron beta decay. For many years M_v was thought to be zero, but in 1998 experiments showed that it is nonzero, though tiny. In addition, two heavier electrons and two heavier neutrinos have been discovered. The masses of the second and third of the three electron partners are different from one another; ditto for the other pair of neutrino partners. The significance of these results for astrophysics is enormous, as I will explain soon. (They are also highly significant for elementary-particle physics and cosmology.)

Let us now resume the story of proton-burning reactions that power stars like the sun. In the first of them, two protons react, producing a deuteron, a positron, a neutrino, and energy to be shared among them:

$$p + p \rightarrow d + e^+ + v.$$

The deuteron (d) is a bound state of a proton and a neutron: d = (n + p), where the parentheses signify a stable nucleus. The p + p process itself is a *reaction*, wherein the initial pair interacts and produces other particles. Arrows thus serve a double purpose: they signify either a decay or a reaction.

Although the p + p reaction may look like a straightforward process, it is not. First, it incorporates the transformation of a proton into a neutron, a positron, and a neutrino. This feature is most easily seen by replacing the deuteron by its n and p constituents, leading to

$$p + p \rightarrow (p + n) + e^+ + v.$$

On carrying out a symbolic "subtraction" of the proton (p) common to both sides of the arrow, the remainder seems to be (the forbidden) proton beta decay, namely

$$p \rightarrow n + e^+ + \nu,$$

where e^+ stands for a positron.

To understand how this seemingly paradoxical result arises, recall that proton beta decay into a neutron is forbidden when the proton is isolated. The presence of two protons in the p + p reaction means that neither of them is isolated. More importantly, the sum of the two proton masses is greater than the sum of the masses of the deuteron, the positron, and the neutrino. Why is this so? Because the deuteron, being a *bound* state of the neutron and proton, is in a quantum state whose total energy is less than $(M_n + M_p) \times c^2$, the latter being the energy of an *unbound* neutron–proton pair. The difference in energies, $[(M_n + M_p) - M_d] \times c^2$, is known as the *binding energy* of the deuteron; it is the extra energy that allows the p + p reaction to be exothermic.

This is the reason, energetically speaking, that the first of the proton-burning reactions occurs. The p + p reaction, however, enjoys another feature that makes it less than straightforward: it is mediated by the weak force. The weak force is involved because the reaction produces a neutrino. This force is so short-ranged that the proton beta decay portion can only occur if the two protons are exceedingly close together. But countering this intimacy requirement is the electrical repulsion between the two protons, which becomes more repulsive as the two protons get closer together (like-sign charges repel one another, whereas opposite-sign charges attract.) Because the electrical or *Coulomb* force acts to keep the protons apart, and the weak force is truly weak, another seeming paradox arises, viz., why doesn't the repulsive Coulomb force simply prevent the reaction from proceeding?

One possibility for resolving this paradox might be the mutual gravitational attraction of the two protons: is it sufficient to overcome their electrical repulsion? The answer is not simply No, it is a resounding NO!

The reason for the strong No is that for all proton–proton separations, the gravitational force is essentially infinitesimal compared with the Coulomb force. The simplest way to show this is to form their ratio. These two forces depend on the proton separation in exactly the same way, so that the separation cancels out after forming the ratio. The quantities that remain are known constants, such as the magnitude of the elementary charge and

Newton's gravitational constant. Inserting their values and carrying out the calculation, the ratio of the gravitational to the Coulomb force is found to be approximately 10^{-36}, a number certainly small enough to qualify as infinitesimal: the attraction of gravity can *never* overcome the Coulomb repulsion.

If the effects of the Coulomb repulsion between protons could not be overcome, the sun would not shine. Because it does, something omitted so far needs to be considered. That something is temperature, and it enters the analysis through the motion of the protons, as indicated on page 64. Since temperature is a measure of the speed of the objects that make up a substance, the consequence for solar (and all stellar) energy generation is that if the temperature is high enough, a pair of energetic protons can come close enough together that they can undergo the peculiar quantum process of *tunneling through* the repulsive Coulomb barrier that acts to keep them apart, thereby initiating the weak-force transformation of a proton into a neutron, a positron, and a neutrino.[11]

The probability for the occurrence of tunneling increases with the temperature. When the center of the sun (which has the highest temperatures and therefore is where the reactions take place) reaches a temperature of about 15 million K, tunneling allows the p + p reaction to proceed. You may wonder at this point if all objects can undergo quantum tunneling. The answer is No: tunneling is an event that requires microscopic masses. For example, the probability that a person leaning against a wall will tunnel through to the other side is vanishingly small. Even the pair of tiny-mass protons do not have an easy time of it: at $T = 15 \times 10^6$ K, and under the pressure and density at the center of the sun, the p + p → d + e^+ + v reaction occurs roughly once per 9 billion years! That it happens constantly is testimony to the gigantic number of protons (roughly 10^{57}) in the sun, enough to keep it steadily "on fire."

Once the p + p tunneling reaction forms deuterons, the next step in the proton-burning chain is the interaction of a deuteron with a proton. It produces a photon plus the nucleus ^{3}He, which contains two protons and one neutron. ^{3}He is an *isotope* of the alpha particle, the two-proton, two-neutron nucleus of the helium atom. (Isotopes of a nucleus have the same number of protons but differing numbers of neutrons.) After ^{3}He is formed, there are three paths by which alpha particles can be created, each occurring with a different probability. What happens overall is that four protons are burned, and alphas, positrons, neutrinos, and photons are generated, an outcome I summarize by

$$4p \rightarrow \alpha + 2e^+ + 2\nu + \text{photons},$$

where α, the Greek letter alpha, denotes the alpha particle, and the arrow now stands for a whole set of reactions, not a decay of the four protons.

These latter processes are *fusion* reactions, in which two nuclei synthesize, or fuse together, forming a new nucleus more massive than either of the initial two. Such reactions are the opposite of fission, wherein one nucleus bombards a second and breaks it up into two or more nuclei. Both processes occur in nuclear bombs; the sun, in fact, is a giant fusion bomb held together by gravity.

The proton-burning chain of events creates two neutrinos plus the photons needed to help counterbalance gravitational collapse. Because the neutrinos interact so weakly with matter, they escape quickly from the sun and readily pass through almost everything they encounter without normally leaving a trace: even as you read this, gigantic numbers of neutrinos are traveling through the earth and your body. Photons, on the other hand, interact far more strongly, being easily scattered by charged particles. To leave the sun they must traverse a distance equal to its radius. On the way they are constantly being jostled by the charged particles they encounter. The distance between such encounters is denoted the *mean free path*, which Phillips (1994) estimates to be about 1 millimeter, leading to an average time of roughly 50,000 years for a photon to reach the sun's surface from its center! The sun is indeed opaque.

Energy Production and the Sun's Lifetime

The age of the sun is about 5 billion years, but in contrast with a person, whose particular life span cannot be predicted in advance, the sun's lifetime *can* be estimated. The estimate is made using the amount of energy produced by the conversion of four protons into an alpha particle. Calculation of that energy involves the same procedure as in neutron beta decay, although now I will put numbers into Einstein's formula.

For energy to be produced, the mass of an alpha particle must be less than that of four protons. The energy released is the difference of these two sets of masses multiplied by c^2: $[(4M_p - M_\alpha) \times c^2]$, where M_α is the mass of the alpha particle. This expression is the *binding energy* of the alpha, in analogy to the *binding energy* of the deuteron, introduced on p. 75. (By the way, the mass energies of the positron and neutrino are small enough compared with the binding energy that they have been ignored. The relevant number is roughly 0.00025 compared to 0.0066.)

It is convenient to express the alpha particle binding energy as a fraction of the mass energy of the four protons. Forming the ratio and inserting the relevant numbers yields

$$(4M_p c^2 - M_\alpha c^2) \,/\, 4M_p c^2 \cong 0.0066.$$

That is, roughly 7% of the mass of the four protons provides the energy of the final photons and neutrinos, with most of it residing in the photons.

The 0.0066 value of the ratio is one of the keys leading to the 10-billion-year estimate of the sun's lifetime. The lifetime is estimated by calculating the total energy that would be available from proton burning and then dividing that energy by the luminosity. The former quantity has dimension energy, the latter energy per time; the division yields the time over which the energy can be released, which is the sun's lifetime. Such a calculation is analogous to determining the time it takes to travel a certain distance at a constant speed: one divides the distance by the speed (which is distance per time) to get the time. For example, it will take a half hour to travel 30 km at a speed of 60 km/hr. In the case of the sun, the available energy is equal to the total number of protons in the sun (about 10^{57}) times the energy released when four are burned (0.0066 enters here) times a factor estimated to be about one tenth, which takes account of the fact that most of the sun's protons will not be burned prior to its becoming a red giant. Evaluating this energy and dividing it by L_{Sun} then leads to the 10-billion-year estimate for the sun's lifetime.

This estimate of the lifetime is consistent with the occurrence time for the reaction p + p → d + e$^+$ + $\bar{\nu}$, for which the sun's central temperature is a critical component. Because the central temperature is higher in more massive stars, the occurrence time for the preceding reaction decreases significantly: it goes down by a factor of about 30,000 when the central temperature is roughly twice that of the sun. As a result of the higher temperature, the luminosity increases (for a blackbody, L varies as the fourth power of T). Because so much more energy is emitted per second, such stars live a much shorter time than a star having the sun's mass.

It follows from proton burning that stars contain protons, alpha particles, photons, and neutrinos. Stars are believed to form from the gravitational collapse of matter consisting mainly of protons; this shrinking in volume of a sufficiently large amount of matter leads to a rise in temperature, which eventually becomes high enough for proton burning to occur. This scenario raises a question that you may have been asking yourself: "Where do the protons come from?" The penultimate answer is that they were synthesized soon after the Big Bang from elementary particles known as quarks and gluons, as I will discuss in the final chapter. But the answer to the ultimate question, concerning the origin of quarks, gluons, the energy of the Big Bang, and indeed of the Big Bang itself, is at present unknown. One reason for this is the existence of an earliest time—approximately 10^{-43} seconds after the Big Bang—before which current theories are inapplicable. In place of them, cosmologists have proposed conjectural frameworks based on conjectural frameworks proposed by elementary-particle physicists; some of them are examined in the final chapter.

Testing the Energetics Theory

The preceding delineation of the constituents and energy source of the sun is part of a general theoretical description of stars. From it stellar models are constructed that yield radial distributions of the temperature, pressure, and density throughout the star, as well as stellar lifetimes, evolutionary tracks, and end stages. These models and the underlying theory are the foundation for the current understanding of stars, and so it is essential that there be no doubts concerning their validity. Although the calculated results are both reasonable and consistent, they alone do not constitute a definitive test of the models. Only one such test has been proposed; for obvious reasons the sun and its theoretical description—denoted the *standard solar model*—is the guinea pig. The test involves detection of the neutrinos produced by the nuclear burning of protons. Success is determined on a pass/fail basis: either the measured number of neutrinos agrees with the theoretical prediction or it does not. A finding of "not" would raise severe questions about the validity of the model.

The neutrino test is very nontrivial to administer: the extreme weakness with which neutrinos interact with matter makes them hard to detect. For example, the existence of neutrinos and antineutrinos was hypothesized in 1930 by the Austrian physicist Wolfgang Pauli but was not confirmed until 1956 by the Americans Clyde Cowan and Frederick Reines (in experiments for which they were awarded the Nobel Prize). However, the difficulty of carrying out measurements has never been an ultimate deterrent to attempting an experiment: not for Cowan and Reines, nor for an American team led by Raymond Davis that began looking for solar neutrinos in 1968.

Davis's detector, which was located deep underground in the Homestake gold mine in Lead, South Dakota, contained 400,000 liters of the cleaning fluid perchloroethylene (C_2Cl_4). It was built underground to minimize false readings initiated by other particles and used C_2Cl_4 because the chlorine (Cl) allows for optimal detection of certain of the neutrinos predicted from proton burning. It is an amusing side note that having ordered so much cleaning fluid, Davis received solicitations from manufacturers of coat hangers!

In 1986, a Japanese team led by Masatoshi Koshiba, using the Kamiokande detector, also began looking for solar neutrinos. This detector, originally constructed in a zinc mine near Tokyo, was designed to search for proton decay (a hypothesized phenomenon described in the last chapter of this book). Both the U.S. and Japanese experiments observed neutrinos, but in each case significantly fewer than predicted by the standard solar model. These results were unsettling, and led to increasingly refined calculations on the standard solar model and the uncertainties associated with it. Although the latter were reduced, the measured discrepancies were not eliminated.

The problem was finally resolved between 1998 and 2002. Its resolution was based on the now-known existence of three types of neutrinos and the fact that each type has a nonzero mass, a circumstance that allows for a new theoretical possibility: the different kinds of neutrino can transform into one another. This behavior, termed *neutrino oscillation*, is an example of an idea first put forward by Bruno Pontecorvo and Vladimir Gribov (an Italian Russian and a Russian, respectively) in 1968, soon after Davis first reported the discrepancy. There are three components to the solution: verifying that neutrinos have mass, that oscillation between different neutrino types does occur, and that the percentage of solar neutrino degradation into the other types is just the amount needed to explain the discrepancies.

Neutrino oscillation was unambiguously demonstrated at an improved Japanese detector called Super-Kamiokande in 1998 by a collaboration of 100 scientists from 23 institutions in Japan and the United States. Four years later, a Canada–U.S. collaboration at the Sudbury Neutrino Observatory located in a mine near Sudbury, Ontario, showed that neutrinos oscillate in just the proper way to account for the discrepancies! This is an exciting and liberating result, providing not only verification of the standard solar model but also renewed confidence that theoretical analysis has indeed led to an understanding of stellar properties, aspects of which are discussed in the next section. The importance of the original undertakings by Davis and Koshiba was given the ultimate scientific recognition in 2002, when the pair shared half of the Nobel Prize in physics (the other half was awarded to Riccardo Giacconi for his pioneering work on X-ray astronomy).

Stellar Nucleosynthesis

The proton-burning chain produces four stable nuclei: the deuteron, the helium isotope ^{3}He, the alpha particle itself, and ^{7}Li, the nucleus containing three protons and four neutrons. Apart from *primordial nucleosynthesis*—the production in the early Universe of these four nuclei plus ^{6}Li—the current paradigm is that stars are the furnaces in which the nuclei of all the other naturally occurring elements are synthesized. Yet at one time, the absence of stable nuclei with total numbers of neutrons and protons equal to 5 or 8 argued against this paradigm. Why? Because these absent nuclei were thought to be the only bridges by which heavier nuclei could be formed from the four created by proton burning. One of these putative bridges is ^{8}Be, the beryllium nucleus consisting of four neutrons and four protons. By absorbing an alpha particle, it could form the carbon nucleus designated ^{12}C. However, ^{8}Be is an unstable nucleus whose lifetime is about 7×10^{-17} sec, too short for it to partake of the normal reactions that would yield ^{12}C.

While carbon (and oxygen) nuclei cannot be formed in the normal way via reactions involving ^{8}Be, they and heavier ones must somehow be created in stars, and the question was how? The answer was provided in 1954 by the English astrophysicist Fred Hoyle, who showed that if a hitherto unknown excited state existed in the stable carbon isotope ^{12}C, then the collision of three alpha particles not only could populate it, but it would also decay to the stable ground state of ^{12}C *via* photon emission. Soon after, the experimental verification of the designated ^{12}C state and its decay modes conclusively established the three-alpha-collision path as the means for creating carbon nuclei in stars (oxygen is formed by fusing carbon and an alpha particle). Because the lifetime of the ^{8}Be nucleus formed in three-alpha collisions is so short, Hoyle's state is produced only when central stellar densities are about 8000 times that of water (or greater).

Hoyle's process is the key to explaining stellar synthesis of the nuclei from carbon up through the iron isotopes. All are formed in *exothermic* (energy-releasing) reactions, which will occur as long as stellar central temperatures are high enough to overcome the Coulomb repulsion that would otherwise keep the nuclei too far apart to initiate reactions.

The iron isotopes are the heaviest that nuclear reactions can create in normal stars: heavier ones are generated only via *endothermic* reactions (those in which energy is absorbed), and unfortunately there is insufficient energy available in normal stars for this type of process to occur. (They are also not created in the aftermath of the Big Bang.) Since heavier nuclei and their corresponding atoms do exist, stars must therefore evolve in ways that eventually allow the necessary endothermic reactions to take place. Two such are the expansion of high-luminosity stars into super giants whose very hot outer envelopes supply the needed energy, and the explosive, supernova phase of less-luminous stars, each of which you will encounter later in this chapter.

Correlations: The Hertzsprung–Russell Diagram

The foregoing summary has described certain properties of and some of the processes that are believed to be taking place in individual stars—with "our" star, the sun being singled out for special attention. I now turn to a different aspect of stars: their behavior as members of stellar populations and an evolving species with specific types of end stages.

The measurable attributes of stars are their spectral types (photospheric chemical compositions), surface temperatures, masses, and luminosities. In the early part of the 20th century, two astronomers, Ejnar Hertzsprung, a Dane, and Henry Norris Russell, an American, decided independently to look for correlations between luminosity (L) and spectral type/surface temperature (T). They began by selecting a sample of nearby stars whose distances from the earth were known, thus ensuring that the luminosities could be estimated. Each man then constructed a graph of luminosity *versus* spectral class/temperature, with the former in the vertical direction and the latter in the horizontal one.

Because L and T were known for all the stars in the sample, each star was represented by a point on the graph, that point being the unique value of the star's luminosity and spectral class/temperature. A correlation would exist if the collection of points fell into patterns of one sort or another, while a more or less random distribution of the points would mean no correlation at all. Such a construction is known as a Hertzsprung–Russell diagram, and beginning with the first ones created, they all have exhibited correlations.

In this current era of quantification, correlations of various kinds are probably familiar to you. One example is the strong cor-

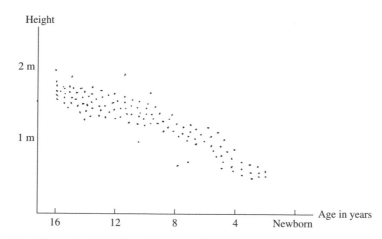

Figure 13. Hypothetical distribution of children's ages *versus* their heights, with the ages shown in the horizontal direction and heights in the vertical. The clustering of the "data" points into a broad band running from lower right to upper left is consistent with the values expected in a normal population of children in developed countries.

relation between income and educational level: typically—but not always—the income of people who have earned professional degrees is greater than for persons without such training.

Another correlation, one that has a "hidden" component and is therefore analogous to the Hertzsprung–Russell diagram, would be seen if a plot of height *versus* age were made for a large sample of children between ages 2 and 16 years.[a] In general, height increases with age, and you should expect to find on such a plot that the height/age combinations for most of the children would be located in a broad band, with relatively little scatter. That is, there should be no 3 year olds who are 2 meters tall (!) and very few if any 15 year olds who arc 1 meter tall, whereas some 12-year-old children might be taller than some who are 16. A hypothetical plot of this particular correlation is shown in Figure 13. Although hypothetical, the underlying idea is not only meaningful, it also serves the purpose of providing a "snapshot" of a population: it gives the viewer a feel for how a typical member of the population would change in time, without the need for following any one of the them for the total time period.

---

[a]Webb (1999) is the inspiration for this example, modeled on one of his.

The latter concept is manifested in the figure by the broad band running from lower right to upper left. There is a small scatter away from the main band of points, each of which represents one child from the hypothetical sample. Overall, the band conforms to our notions of growth: as a child grows older, its height increases. The hidden component referred to above concerns weight: the broad band is also a measure of the weights, or equivalently the masses, of the children it represents, for as a child ages, it normally becomes heavier. In fact, there is a three-way correlation among age, height, and weight, but the weight portion is roughly inferable from the age/height correlation. A similar situation arises in the case of the Hertzsprung–Russell (H-R) diagram.

H-R diagrams are normally created from a population of stars in close proximity to one another, for example, those forming a *globular cluster* (a type of stellar system defined in Chapter 5). As with Figure 13, the H-R diagram provides a portrait of a stellar population over both mass and time, the latter indicating how stars in the sample evolve.

Figure 14 is a schematic depiction of the diagram, with the spectral class letters specified on the upper horizontal line and surface temperature T indicated on the lower one. In it, T increases from right to left, just as age does in Figure 13. The luminosity L, measured in units of the sun's luminosity L_{Sun}, is designated along the left-hand vertical line. Each star is represented by a single point, and four patterns (correlations) are evident. I shall concentrate on the broad, roughly diagonal swath, running from the lower right to the upper left, along which most of the stars fall; it is denoted the *Main Sequence*. Relatively few stars are at its extreme upper end, as this portion corresponds to very hot, shorter-lived stars. The sun appears on the Main Sequence as the small diamond at the values unity on the vertical scale and G2/5780 K on the horizontal ones.

Stars are formed from massive amounts of gravitationally collapsing clouds of matter that, if the mass is large enough—greater than about $0.08 M_{Sun}$—will eventually reach sufficient internal temperatures to begin burning protons. It is at this point in the evolution of sufficiently massive stars that they move onto the Main Sequence (MS), remaining there for either all or a majority

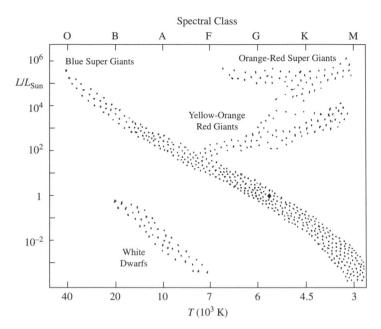

Figure 14. Schematic representation of a Hertzsprung–Russell diagram. The lower horizontal axis is the absolute temperature measured in units of $10^3 K$, and the upper horizontal axis displays the spectral classes, with the letters defined in the text. On the vertical axis is the luminosity, given in units of the sun's luminosity, L_{Sun}. The broad band running along the diagonal from upper left to lower right is known as the *Main Sequence* and contains the majority of the stars in the population contributing to the diagram. The other sets of stars in it, each containing relatively few members, just as with the heavily populated Main Sequence, is typical in terms of relative numbers and placement.

of their lifetimes, generating energy through nuclear fusion. For about 80–90% of them, this occurs by the proton-burning reactions described previously. Most of the other stars generate energy by the so-called carbon–oxygen cycle, which not only produces photons and neutrinos but also luminosities that depend more strongly on the surface temperature than for the proton-burning-reaction stars. A description of the carbon–oxygen cycle that includes technical details can be found in Phillips (1994).

The hidden component of the H-R diagram is mass, just as in the example of the putative correlation for children delineated in Figure 13. Mass increases from right to left in Figure 14; a rough approximation is that L is proportional to the third power of the mass M $(L \propto M^3)$. Because greater luminosity means shorter lifetimes, then an increase in stellar mass leads to a decrease in lifetime. Very hot, high-luminosity stars are the most massive and short-lived: a very hot blue star with a mass about $10 M_{Sun}$ and a luminosity about $10^4 \times L_{Sun}$ is calculated to remain on the MS for about 10^7 years. At the other extreme, cool, low-luminosity stars are the least massive and most long-lived: a star with a mass about $0.2 M_{Sun}$ and a luminosity of $10^{-2} \times L_{Sun}$ is predicted to live for approximately 500 billion years on the Main Sequence.

Stellar Masses

Mass plays an extraordinarily influential role in stellar behavior: it is *the* controlling factor—almost everything depends on mass. For instance, mass determines whether a collapsing gas cloud becomes a planet, a normal star, or a brown dwarf star (which is almost too faint to be seen); what a star's central temperature will be and therefore whether it generates energy via the proton–proton or the carbon–oxygen cycles; how luminous a star will be; how long a star will remain on the Main Sequence; and what the evolution off the MS as well as the end stage will be.

As implied by the H-R diagram and the way L scales with M, you might expect that there are upper and lower limits on stellar masses. A theoretical lower limit is roughly $0.08 M_{Sun}$, based on the minimum mass needed to initiate proton burning. This limit is respected by observation: the smallest stellar mass is about $0.1 M_{Sun}$. Theory also suggests an upper limit of about $100 M_{Sun}$, based on the role played by radiation pressure: if M is too large, the star's radiation pressure will be too great, causing a hydrostatic instability. This provokes destabilization and a consequent eruption of matter, leading to a smaller-mass remnant. Consistent with this idea, the maximum stellar mass observed is about $50 M_{Sun}$.

Accurate values of a star's mass can be calculated if the star is a member of a binary pair, whereas the mass of an isolated star can be determined only in special circumstances. Once the binary character is established, theory enters the calculation through the generally valid assumption that Newtonian gravity suffices to describe the motion of the pair. H-R diagrams are usually constructed for nearby stars that occur in pairs and thus have known masses. However, the diagram is a well-enough understood tool that if Ls and surface Ts can be measured for a new sample of stars in proximity to one another, the resulting H-R diagram can be used to estimate the masses with fair accuracy. Furthermore, the Main Sequence as well as the tip of the red giant portion of an H-R diagram can serve as a step in the cosmic distance ladder.

A star's mass governs whether it will remain on the Main Sequence burning protons or, after enough of its fuel has been burned, move off it and evolve into another stage, such as a red giant or super red giant. These stages appear in the H-R diagram of Figure 14 as the pair of patterns on the upper right. Betelgeuse, located in the constellation Orion, is a red giant, and billions of years in the future, the sun will become one.

For stars with masses close to the sun's, the red-giant phase lasts about a billion years, ending with an explosive ejection of much of its matter, whose interactions with charged particles turns it into a glowing *planetary nebula*. The remaining matter becomes a *white dwarf* (they are found in the lower left portion of Figure 14). An instance of a white dwarf is the faint *companion* of Sirius, the brightest star in the sky. These two constitute a binary pair, with the star designated as Sirius A, and the faint dwarf as Sirius B. Located in the constellation Canis Major, they are about 2.64 parsecs away.

For stars with masses greater than roughly $6M_{Sun}$, evolution off the Main Sequence occurs via a supernova explosion. The star's initial mass determines if the remainder becomes a white dwarf, a neutron star, or a black hole, the three end-stages of stellar evolution. Whether a star remains on the Main Sequence or not, most will eventually use up their fuel, cease radiating, and become cold balls of nonradiating matter known as black dwarfs. Only those that become black holes can escape this fate: instead, they can end by evaporating all their mass via a process explained at the end of this chapter.

Evolution Off the Main Sequence: Cepheid Variables and Red Giants

Stars evolve *onto* the Main Sequence, their time on it being a stage in their evolutionary journey. Different mass stars will leave the MS via different routes. Of major significance for astronomy and cosmology is the evolution of some stars into the stage known as a *Cepheid variable*. A *variable* star displays periodic changes in its luminosity; in Cepheids the changes result from radial pulsations. In other words, Cepheids expand and contract, as in breathing. For example, the radius of δ Cepheus ranges between 23 and 26 solar radii; its maximum luminosity is about twice that of the sun. Most Cepheids are yellow giants, with luminosities in the range 10^3 to $10^4 \times L_{Sun}$.

The name *Cepheid* comes from the constellation Cepheus, whose star δ (delta) Cepheus was the first of this type to be observed.[12] Another is Polaris, the pole star. Stars with masses between about five and twenty times the sun's mass become Cepheids as they evolve into the red-giant phase.

As a Cepheid's photosphere periodically pulsates inward and outward, the changes in luminosity are correlated with changes in the Doppler shifts of its spectral lines. Because the period P of the pulsation is also related to the Doppler shifts, there is a correlation between L and P. In 1912, the American astronomer Henrietta Swan Leavitt announced (in a Harvard College Observatory Circular written by her boss!) a seminal discovery: not only is there a correlation between L and P, they are also related mathematically.

Leavitt found that L is related to P by a formula containing certain constants. Once the constants are determined, then measurement of P uniquely determines L, the intrinsic luminosity of the star. Her discovery not only provides intrinsic luminosities (L), it also led to a new method for determining cosmic distances. Recall that astronomers measure ℓ, the *apparent* luminosity of the star, a quantity smaller than the intrinsic luminosity by the square of the star's distance D away ($\ell \propto L/D^2$). If either L or D can be measured separately, the value of the other follows from a measurement of the apparent luminosity, but if only ℓ is measured, neither L nor D can be evaluated. Leavitt's period/intrinsic-luminosity relation solves this problem for Cepheids, most of whose distances were not known: measurement of P yields the value of L, and knowledge of L coupled with measurement of ℓ determines D. In other words, Cepheids are distance indicators! This remarkable result turned out to be the key to galactic distances and thus to the expansion of the Universe.

Cepheids occur as an evolutionary step of a MS star toward its red-giant phase, a stellar stage in which the increase in radius is huge. For example, when the sun becomes a red giant, its radius is predicted to grow by a factor of about 70 or more, thereby engulfing Mercury and cooking the surfaces of Venus and the earth. A billion years or so after that, the sun will become a red super giant, its envelope will expand to engulf Jupiter and, after other events recounted below, it will finally become a white dwarf. Not to worry, of course: the standard solar model predicts that this will not take place for billions of years.

The red-giant phase is predicted to occur for most Main Sequence stars with masses greater than or equal to M_{Sun}. (In contrast, lower-mass stars do not evolve into other stages; they remain on the MS for tens of billions of years until they burn out and become black dwarfs.) The red-giant phase will be entered when about 80% of the protons in the central proton-burning core have fused into helium, at which point proton burning ceases. With too few protons remaining to supply the needed internal pressure, the core will contract under gravity and the central temperature will rise, causing the star to move off the MS.

Two new types of burning then start to occur: the alpha particles in the core fuse to form carbon nuclei, and the protons in the region outside the core fuse to produce alphas. The release of energy in the latter process leads to the huge expansion in radius that sends the star into its red-giant stage. With so much more surface, the star's brightness increases, even though its surface temperature actually decreases. An example is Betelgeuse, whose radius and luminosity, respectively, are about 350 and 12,000 times that of the sun, while its effective temperature is about half the sun's.

After the helium-burning phase, the resulting increase in the star's radiation pressure ejects its outer layers, following which the continuing irradiation ionizes many of its atoms. The ejected outer layers then become a brightly colored, glowing cloud of gas known as a *planetary nebula*, beautiful examples of which can be found, for example, on the Hubble Space Telescope Web site listed in the Bibliography. The remaining material forms a tiny, high-temperature core—a star—whose mass I denote by M_{core}.

The value of M_{core} now becomes crucial, as it alone determines the end stage of the star's evolution. Theory leads to the following conclusions: if M_{core} is not greater than about $1.4M_{Sun}$ (the so-called Chandrasekhar limit, named for the Indian-American astrophysicist who derived it), then the remainder becomes a white dwarf; if the core mass lies between $1.4M_{Sun}$ and approximately $3M_{Sun}$, the remainder will become a neutron star[13] (rotating neutron stars are called *pulsars*, which emit radio waves in a searchlight-like beam); for M_{core} between $3M_{Sun}$ and $5M_{Sun}$, it probably will become a black hole, while for larger masses it is almost certain to become one. End stages are the subject of the next three sections. My treatment is qualitative, but well-prepared readers who wish to learn more of the quantitative aspects are recommended to the text by Phillips (1994), while those of you who seek more qualitative information on black holes and the wonders associated with them might try the popular book by Thorn (1994).

White Dwarfs, Type II Supernovas, and Neutron Stars

In addition to the correlation between M_{core} and the different end stages, theory predicts a relation between the initial stellar mass and M_{core}, or equivalently (and approximately), it specifies which stars will become white dwarfs, neutron stars, or black holes. Letting M represent the star's initial mass, the results are as follows: for M less than about $6M_{Sun}$, the end stage will be a white dwarf; for M between 6 and about 25 times the sun's mass (or slightly more), the end stage will be a neutron star; and for M greater than about $30M_{Sun}$, the end stage will be a black hole.

The name *white dwarf* is aptly chosen. These stars are white because their surface temperatures are high, as indicated by their positions on the Hertzsprung–Russell diagram; equivalently, they are members of spectral classes A or B. Despite surface temperatures that can reach nearly 10^5 K, white dwarfs are dim, not bright. The reason for this resides in the second part of their name. Being dwarf stars, they are so small that their surface area is tiny, as compared, for example, with that of the sun. With so little area to supply photons, their luminosity is low, and thus they appear dim (Sirius completely outshines its white dwarf companion).

Just how small are white dwarfs? The answer is based on extremely well founded theoretical analyses, ones that date from Chandrasekhar's original investigation. Not surprisingly, the radius of a white dwarf depends on its mass, but a reasonable order of magnitude estimate for it is 6400 km, the radius of the earth (and roughly one-hundredth of the sun's radius). The radii of the white dwarfs observed in binary situations are consistent with this rough estimate [Phillips (1994)]. Compared with the sun, the typical surface area and volume of a white dwarf are smaller by approximately 10,000 and a million, respectively.

At this point you may be wondering why a nonburning solar mass contained in one-millionth of a solar volume doesn't collapse under its own gravity. In other words, why do white dwarfs exist at all? The answer to this perfectly reasonable question lies in the quantum nature of white dwarfs. Prior to becoming a dwarf, the core is mostly carbon *nuclei*, but in the process of becoming a dwarf the nuclei capture electrons, an event that turns the carbon nuclei into carbon *atoms*. The electrons in these newly-formed carbon atoms oppose the gravitational collapse, and they do so because their behavior is governed by quantum theory.

Electrons belong to a class of objects known as *fermions*, named for the great Italian-American physicist Enrico Fermi. Quantum theory mandates that no two quantum states in a system of identical fermions—such as the electrons in a white dwarf—may be the same. In a white dwarf, the electrons are in the lowest possible energy states available, a situation known as *electron degeneracy*. If gravitational collapse could occur, some of the electrons would be forced into states already occupied by other electrons, but as this is forbidden by their fermion nature, the star does not collapse: electron degeneracy acts as a counter pressure[b].

White dwarfs are macroscopic examples of a quantum system once thought to exist only at the microscopic level. Electron degeneracy leads to a white dwarf density d_{WhDw} approximately equal to $10^{10} \times d_{\text{sun}}$, as stated in Table 7 (see Chapter 2). For $M_{\text{core}} = 0.4 M_{\text{Sun}}$, the values of surface temperature and luminosity are calculated to be 10^4K and $10^{-3} \times L_{\text{Sun}}$, numbers consistent with the comments made above.

White dwarfs exist because of degeneracy pressure, but this pressure will fail to counter the collapse when M_{core} exceeds the Chandrasekhar limit of $1.4 M_{\text{Sun}}$. In this case, the effect of the added gravity is to create neutrons by forcing the electrons to combine with the protons in the nuclei of the carbon atoms. (Since the lowest states are occupied in a white dwarf, this is the only place the electrons can go.) The addition of the new neutrons to the existing ones transforms the core into a tiny, gravitationally bound gas of neutrons. What might have been a white dwarf now becomes a *neutron star*.

For stellar masses greater than about $8 M_{\text{Sun}}$, neutron star and black hole end-stages are reached *via* the phenomenon known as a *type II supernova*. The term *nova* was coined long ago for a star that suddenly appeared in the sky (nova means "new" in Latin). Novas are now believed to be an erupting event on the surface of a white dwarf when it accretes material from a red-giant companion. Nuclear reactions then occur on the surface, causing the dwarf's luminosity to increase by factors of ten to many thousands, so that to the naked eye it seems as if a brand new star has appeared.

Supernovas are exactly what the word implies: they are super or gigantic novas, events in which a star *explodes*, suddenly

---

[b]Chandrasekhar deduced the role of electron degeneracy on the ship that took him from India to England in the early 1930s. In one of those odd quirks of scientific failures of understanding, his work was denigrated at a meeting by A. S. Eddington, the outstanding astrophysicist of his time. Eddington never admitted his error, but Chandrasekhar enjoyed a highly distinguished career in astrophysics and cosmology.[14]

increasing its luminosity to as much as $10^8 \times L_{Sun}$ and ejecting significant amounts of stellar material. While their occurrence is normally a rare event, the huge number of stars in a galaxy should dramatically increase their frequency of occurrence. Webb (1999), for example, notes that one should occur in the Milky Way as often as every 50 years or so. However, the presence of very large quantities of obscuring dust has meant that in the past millennium, only five of them have been seen.

As implied by the phrase *type II*, there is another class of supernovas, known as (what else?!) type I; each of the two classes is divided into several subclasses. Type Ia, which serves as another very important distance indicator, is examined in the next chapter. Type IIs are believed to arise from the following sequence of events in a relatively young, sufficiently massive star. During its time on the Main Sequence, it will have created nuclei up to and including iron. Structurally, its interior will resemble an onion: surrounding an inner core consisting of iron nuclei will be a series of shell-like layers containing first the nuclei of silicon (next to the iron core), then magnesium, and so on to the progressively lighter nuclei until the outermost shell of protons is reached.

An onion is a reasonably stable body: it will last for many days, even unrefrigerated, before it begins to decay. The onion-like structure described above, however, does not share this characteristic: once the star has generated its iron core, it becomes highly unstable. Nuclear reactions will have ceased, because insufficient energy is available to power what would be endothermic reactions. The absence of nuclear reactions leads to a pressure drop in the iron core, which can then no longer resist the force of gravity, and the iron core collapses. As it does, a deluge of high-energy photons is produced that rapidly disintegrates the iron nuclei into their constituent neutrons and protons.

While the once-iron core continues to collapse, its newly released protons absorb electrons, thereby producing neutrinos and neutrons. The latter form a core of neutrons. Meanwhile, the material that surrounds the collapsing core is also contracting, and the collision of the two collapsing portions leads to the supernova explosion, wherein endothermic nuclear reactions occur that create elements heavier than iron. In addition, more neutrinos are produced, and as much as half or more of the star's mass is ejected, eventually forming a planetary nebula.

Supernovas lead to the destruction of the exploding star. Even though the star is dying, its luminosity can become as great as that of a billion stars—which means it shines as brightly as a galaxy. This enormous outpouring of energy can last as long as a month. The huge numbers of neutrinos produced are observable, at least in principle; the first such were measured in 1987 by the Kamiokande detector (mentioned in connection with the measurement of solar neutrinos). These neutrinos came from supernova SN1987A, which exploded in the Large Magellanic Cloud, a neighboring galaxy. Their measurement strikingly confirmed the theory describing this type of cataclysmic event.

While the focus so far has been on the energy associated with the cataclysm, the internal event is the formation of the neutron core. For an initial stellar mass in the rough range of six to about twenty or twenty-five solar masses, the neutron core becomes a neutron star. A rotating neutron star is called a *pulsar*; one, located in the Crab Nebula (the remains of the supernova of 1054, recorded by Chinese astronomers), is a copious source of X-rays and has a luminosity approximately equal to $4.6 \times 10^{31}\,\mathrm{W} \cong 10^5 \times L_{\mathrm{Sun}}$. Theory leads to a radius for this object of approximately $10\,\mathrm{km}$, and a density $d_{\mathrm{Neutron\ star}} \cong 10^{15} \times d_{\mathrm{Sun}}$ (Table 7). This is an enormous amount of matter to stuff into such a small, zero-pressure volume, and the same considerations concerning degeneracy pressure in white dwarfs apply to neutron stars.

That is, because neutrons are also fermions, no two of them can be in the same quantum state. Since they occupy the lowest set of states (or energy levels) available to them—there are about 4×10^{57} neutrons in a neutron star—they give rise to a degeneracy pressure ensuring stability, just as long as the approximate upper limit of $3M_{\mathrm{Sun}}$ on the mass of the neutron core is not exceeded (see note 13). For masses greater than this, gravity again becomes too strong for the degeneracy pressure to resist, and the neutron core collapses even further, now becoming a black hole, probably the best known of all the astronomical/cosmological objects.

Black Holes

The concept of a black hole is based on an Einsteinian gravity phenomenon first studied by the German astrophysicist Karl Schwarzschild. While serving in the German army during the First World War, he not only read Einstein's paper on general relativity, he also used it to analyze the effects of gravity due to a stationary, idealized *point mass* (i.e., a mass located at a single point in space, rather than occupying a finite volume; the point-mass electron is an example). It is remarkable that a person on active duty at the Russian front would have been able to study Einstein's paper. It is even more remarkable that Schwarzschild was able to write two papers detailing his investigations, which he then sent to Einstein. Einstein presented them to the Prussian Academy of Sciences in Berlin, which later published them. Although Schwarzschild's work eventually led to a paradigm shift in the understanding of cosmic phenomena, he contracted a fatal illness at the front and died without seeing the fruits of his labors in print.

Schwarzschild discovered that the gravitational effect of a point mass is to distort or warp the space surrounding it in a fascinatingly peculiar fashion: that space is divided into two distinct portions such that the behavior of radiation or matter in one differs enormously from its behavior in the other.[15] These two portions are the interior and the exterior of a sphere known as the *event horizon*. The radius of this sphere—centered on the point mass—is denoted the *Schwarzschild radius* D_{Sch}, and like so much else in astrophysics and cosmology, D_{Sch} depends only on the value of the mass (apart from certain constants such as the speed of light).[16] The generation of these inner and outer volumes separated by a distinct boundary surface leads directly to the definition of a black hole. It is the remnant of a sufficiently massive star whose core has collapsed under gravity into a volume that is smaller than the event-horizon sphere its mass would have produced were it located at a single point. The collapsed star continues shrinking, all the while living inside its own event horizon.[17]

Black holes are theoretical constructs, based on Einsteinian gravity. An obvious question is whether there are any observable objects in the Universe that can be identified as black holes: do they exist or are they simply wonderful figments of theory—or of a theorist's imagination? The answer to this question involves the bizarre properties of black holes.

Black holes produce gravitational effects that depend on whether they occur inside or outside the event horizon. Suppose first that radiation is external to the collapsed star but inside the event horizon. In this situation, it cannot escape: the black hole's gravity is too strong, and the wavelength of the radiation becomes elongated to such an extent that the redshift defined in the preceding chapter becomes infinite. As a result, the radiation effectively vanishes (infinite wavelength means zero frequency and thus zero energy). Consequently, the collapsed core is invisible: it has disappeared inside its event horizon.

Radiation or mass external to the event horizon will also be influenced by the black hole's formidable gravity. Suppose that matter is falling directly toward the black hole. To an external observer, the speed of the mass becomes slower and slower as it approaches closer and closer to the event horizon, eventually reaching zero at the boundary. This result is a reflection of the fact that in Einsteinian gravity, space is distorted and time is altered. Were an idealized, infinitely long-lived observer to exist outside the boundary, he or she would find that an *infinite* time had elapsed before the mass reached the event horizon. But, for an observer travelling *with* the mass—or, say, for an astronaut who unfortunately happened to be caught by the hole's gravity—the measured time to reach the event horizon would be finite.[18] The mass would continue past the event horizon, becoming invisible as it does so, and would finally experience such extreme elongation on the journey to the collapsed core that it would be pulled apart and disintegrate [see, e.g., Thorne (1994)].

Next, assume that a star is in orbit around a black hole. The hole's very strong gravity would continually pull mass off the star. Such mass (or gas near the hole) tends to form an *accretion disk* that swirls around the event horizon and eventually into it. Compression of the matter in the inner part of the disk heats it, reaching temperatures possibly in excess of 10^6 K. Such high temperatures would lead to the emission of X-rays that in principle could be detected. Observation of such energy is the means to detect an invisible black hole: its existence is inferred by the effect it has on gas or a visible object.

However, X-rays can also be produced when an ordinary star loses mass to a neutron star, and to ensure that the invisible object is a black hole, it is necessary that the star's companion have a mass greater than $3M_{Sun}$, the approximate upper limit for a neutron star's mass (see note 13). It is now widely accepted that the very compact object in the X-ray emitter Cygnus X-1, the first candidate proposed as a black hole, is one. Its mass is about $8M_{Sun}$, well above the required minimum mass limit. Many other candidates are believed to be black holes, of which three different types are thought to exist. These are *stellar* black holes, *supermassive* black holes, and the newest addition to the group, the *mid mass* or intermediate-size black holes.

Stellar and supermassive black holes are at opposite ends of the black hole mass spectrum. The masses of stellar black holes are roughly a few tens of a solar mass; they are in a mutual orbit with a single, X-ray emitting star, as in the case of Cygnus X-1. The Schwarzschild radius of a stellar black hole mass of $10M_{Sun}$ is 30 km, and for each solar mass that accretes onto it, its event horizon radius increases by about 3 km.

In contrast, supermassive black holes, which are believed to exist in the centers of many galaxies, including the Milky Way, have masses in the range of a million to a billion times M_{Sun}; their corresponding Schwarzschild radii are in the range 3×10^6 km to 3×10^9 km. Such black holes, known as *active galactic nuclei*, have grown this large by accreting matter, over a span of millions of years, from many of the stars in the galaxy. Supermassive black holes have also been proposed as the compact energy source of objects known as *quasars*, emitters of X-rays and radio waves with luminosities a hundred times greater than that of galaxies at the same distance from the earth. (Quasars are encountered again in the next chapter.)

Finally, the first existence of a mid mass black hole was deduced in 2002: it was found in the galaxy known as M82, has a mass somewhat greater than $500M_{Sun}$, and is located in a cluster of stars well away from the center.

Evaporation of Black Holes

I commented earlier in this chapter that black holes end their lives not as black dwarfs but by evaporation of their mass. This may seem like an extravagant claim, as their enormous gravity would seem to allow nothing to emerge from inside the event horizon. However, the description of black holes presented so far is based on Einsteinian gravity, whereas the feature that leads to the evaporation of mass is a quantum phenomenon known as *Hawking radiation*, named for the English physicist Stephen Hawking, who proposed the idea.

Hawking radiation is an end-stage mechanism whose origin can be thought of in several ways, for example as arising from quantum fuzziness either in the event horizon or the black hole's gravity. The radiation can take the form of photons or particles, which are created when a tiny amount of the hole's gigantic gravitational energy is used to produce a particle–antiparticle pair or a pair of photons (recall that the photon is its own antiparticle). One of the particles will be sucked into the black hole, the other can escape by the quantum process of tunneling through the attractive barrier that the hole's gravity represents, in analogy to the two protons that initiate proton burning by tunneling through the Coulomb barrier. If photons are created, both can escape.

There is thus a spectrum of possible Hawking radiation. Which particles can be created depends on the size of the hole: the particle cannot be larger than the hole, so that only "mini" black holes can evaporate by creating electron–positron (or proton–antiproton) pairs. In the case of photons, it is their wavelength that cannot exceed the size of the hole, so that the larger the hole, the lower the photon energy.

The probability for these occurrences is very small, and because the decrease in gravitational energy is tiny, the lifetime for ultimate demise by evaporation is immense: that is, it takes a gigantic succession of events to lead to complete evaporation. As with so many things stellar, it depends sensitively on the mass of the black hole: the smaller the mass, the shorter the lifetime (in particular, if M is the mass of the black hole, its lifetime against decay by evaporation is proportional to M^3).

As an example, a black hole of mass equal to $10M_{Sun}$ would live about 10^{65} years, presumably long enough for at least one to be observed. Such a huge lifetime is roughly 10^{56} times the age of the Universe, so that only if the expansion were to occur almost indefinitely would a black hole fully evaporate. On the other hand, if some were created in the Big Bang—and it has been speculated that some, called *primordial* black holes, were so created—none with mass less that about 10^{13} kg would have survived to the present day.[19]

5. The Expanding Universe

Nowadays, anyone wishing to learn what the structure of the Galaxy is can find out from a variety of sources—including this chapter. But, from the late 19th century until 1924, the then-unknown size of the Galaxy was at the heart of a controversy. The controversy concerned the existence of galaxies other than our own: were the so-called spiral and elliptical nebulae[a] in fact other galaxies—"island universes," as the German philosopher Immanuel Kant presciently suggested in 1755—or was the Milky Way so large that it encompassed everything? The American astronomer Edwin Hubble is credited not only with settling the controversy but also with providing the first observational evidence that the Universe is expanding. *Expansion* means that the vast majority of the galaxies are rushing away—receding—from one another. That the Universe is expanding is not simply one of its salient features; it is among the most momentous cosmological findings of the 20th century. The discovery and the interpretation of the expansion of the Universe occurred along a pair of linked paths, one observational, the other theoretical. It is a marvelous story to relate, containing not only controversy, but speculation, error, leaps of the imagination, overlooked results, triumphs, and paradigm shifts; in short, a very human tale, not what is sometimes thought of as the linear progress of science.

Although receding galaxies help make the Universe an active environment, its entire contents are in constant motion, and unlike the chaotic quality of the earth's jiggling atmosphere, these movements are generally orderly as well as measurable. The nearest example of this activity is just underfoot, viz., the earth, which moves in a variety of ways, the most familiar being the rota-

---

[a]*Nebula* is the Latin word for "cloud" and was used to refer to the cloudy patches in the night sky visible to the unaided eye. That some of them can be spiral or elliptical in shape, and indeed are other galaxies, is part of the story told in the present chapter.

tion on its axis at roughly 0.9 km/sec and its revolution around the sun at an average speed of approximately 30 km/sec. In addition to these everyday occurrences are those unnoticed movements of the earth that accompany the motions of the sun.

Relative to the nearby stars, the sun moves with what is known as a "peculiar" velocity; it also revolves around the center of the Galaxy; and it partakes of the overall motion of the 30 or so members of the *Local Group of galaxies*—of which the Large and Small Magellanic Clouds and the galaxy Andromeda are members—toward the Virgo cluster of galaxies, as well as the motion of the local *supercluster* of galaxies toward the Hydra Centaurus supercluster. These latter solar activities end up imparting a speed of about 600 km/sec to the earth.[1] Hence, when estimates of stellar or galactic speeds are made using Doppler-shifted spectra, the various motions of the earth must be corrected for, since otherwise they could influence conclusions concerning distances, a subject I revisit in this chapter. These corrections are always assumed to have been made when relevant numbers are quoted later in this chapter; readers interested in some of the details and general discussions of this topic could consult Webb (1999) or Harrison (2000) and references cited therein.

Astronomers in the second half of the 19th century realized from Doppler-shifted spectra that many of the astronomical emitters of radiation were in motion, the majority receding rather than approaching. But this evidence alone was not taken as an indication of the universal expansion, especially because the prevailing paradigm was that of a *static* Universe. Such a paradigm allows for internal motions, but overall, it yields an unchanging picture. Even the theorists who might be counted among the early cosmologists, such as Albert Einstein, subscribed to this picture, and it profoundly influenced how they thought about their results. Nonetheless, the eventual interpretation of the observations as an *expansion of space itself* was the result of theoretical investigations. The link between the observational and the theoretical paths that led to this interpretation came in 1929, when Hubble discovered an empirical relation between recession speed and distance and noted that it might be a manifestation of the so-called de Sitter effect, which arises from theory.

To have reached his conclusion about the speed–distance relation and also to have been able to resolve a controversy concerning the nature of the Galaxy, Hubble needed reliable distance indicators, ones that could measure distances well beyond the limits of parallax. My description of these new steps on the distance ladder is based in part on material from the past two chapters. After surveying the observational scene and Hubble's achievements, I shall explore aspects of the theoretical interpretation, which is based on Einsteinian gravity, that is, on general relativity.

The Shape and Size of Our Galaxy

In 1610 Galileo, using telescopes he had constructed, discovered that the nebulae now denoted the Milky Way Galaxy and the galaxy Andromeda contained individual stars. He also concluded that nearly all of the other nebulae consisted of stars that were too far away for his telescopes to resolve individually. This problem persisted for more than 200 years: decades after the first parallaxes were obtained (in the 1840s, recall Chapter 2), astronomers still had trouble resolving individual stars in nebulae. It was only through the use of spectroscopy that they first determined the composition of many of the nebulae, some of which were known to be either spiral or elliptical in shape.

Correspondingly, the lack of trustworthy distance indicators made it impossible to draw reliable quantitative conclusions about the shape or size of the Galaxy. Methods such as those of statistical and secular parallax, of moving clusters, and eventually Main Sequence fitting—which went well beyond the trigonometric parallax method of Chapter 2 and are nicely described by Webb (1999)—nevertheless reached out insufficiently far to provide the desired answers. (These methods provided distance estimates not greater than about 500 pc, which is only a few percent of the Galaxy's diameter.)

The person who played the decisive role in determining the shape of the Galaxy was the American astronomer Harlow Shapley. Working in the second decade of the 20th century with the 100-inch telescope at Mount Wilson in California—then the world's largest—he constructed a Galactic distance ladder. Two examples of the important steps in his ladder are Cepheid variables and bright stars in globular clusters (which are gravitationally bound aggregates of as many as a million stars). While his ladder was a shaky one, it turned out to be suitable for the purpose of obtaining the shape although not the correct size of the Galaxy.

Shapley concluded that the Galaxy is a relatively flat disk (like a lens, just as had been claimed as early as the mid-18th century) containing a massive central bulge surrounded by a "halo" of globular clusters. The latter feature was controversial, as was his estimate of the sizes: 90 kpc in diameter, 9 kpc thick at the central bulge, and 15 kpc from the center to the sun. By 1920, Shapley not only had adduced the Galaxy's shape, he was also one of the two protagonists in a "debate," sponsored by the U.S. National Academy of Sciences, whose topic was the nature of the spiral and elliptical nebulae.

This debate, whose other protagonist was the American astronomer Heber Curtis, has become famous in the annals of astronomy, partly because the arguments of two of the preeminent astronomers of the time failed to settle the issue. The reason is simple: neither man had the data necessary to resolve the controversy. In particular, no one then knew how large the Galaxy actually was, despite Shapley's claims to the contrary. Shapley, obviously believing his own estimates of the Galaxy's size, thought it contained everything, including the nebulae; Curtis, on the other hand, disputed Shapley's inferences on the size, arguing for a smaller Galaxy, such that the nebulae were external to it.[2]

Hubble finally resolved the controversy by showing that certain of the spiral and elliptical nebulae at issue *were* galaxies external to our own,[3] just as Kant had speculated. Hubble's investigations, carried out during the period 1919–1924, established the externality of several of the nebulae by using new Cepheid variables he had found with the 100-inch Mount Wilson telescope (the same one that Shapley and many other astronomers had worked with). The distances for three of the nebulae that Hubble estimated from the Cepheid period–luminosity relation conclusively established that they were located well outside Shapley's Galactic limits.

Not only did Hubble's results increase the size of the then-known Universe, they also "destroyed" Shapley's universe (see note 2). Nevertheless, the structure Shapley had deduced for the Galaxy's shape was correct in its essentials; only his size estimates were wrong. They were wrong because of the presence of interstellar dust. The effect of the dust is to absorb starlight, thus rendering the light less intense. Lesser intensity normally means greater distance away, and since Shapley was unaware of the dust's existence, he concluded that the Galaxy was much bigger than it is now known to be.

In the modern picture of the Galaxy, the lenslike portion—two fried eggs back to back[4]—has a diameter of about 30 rather than 90 kpc, the central bulge is approximately 5 and not 9 kpc thick, and the sun is roughly 8.5 and not 15 kpc from the galactic center.[5] The other controversial aspect of Shapley's structure, viz., the existence of a halo surrounding the Galaxy, is not only correct, it is now known to contain a very large amount of nonluminous dark matter, whose composition is unidentified. Dark matter turns out to be an essential ingredient in the Universe, and in later chapters I shall examine evidence for its existence as well as conjectures concerning its nature.

Because the sun is located about half-way out to the "edge" of, as well as being in, the disk, astronomers cannot actually see what the Galaxy looks like: its detailed structure must be inferred—and of course it has been. The disk contains dust plus gas in which new stars are forming, particularly in its spiral arms (the Milky Way is a spiral galaxy). The disk is also the location of young and intermediate age stars, whereas the older ones are in globular clusters, containing up to 10^6 stars, which are found mainly in the Galaxy's halo, where the dark matter resides.

A schematic representation of the main features of the Galaxy as it would appear if viewed from the side and from above (or below) is depicted in Figure 15. However, no schematic drawing can do justice to existing color representations of the Galaxy, and so I draw your attention to Plate 2, which shows galaxy NGC 3949, a "cousin" of the Milky Way. Interested readers are directed to the Hubble Space Telescope Web site for photos of M31 (the galaxy Andromeda), which is also believed to resemble the Milky Way, and to the Cosmic Background Explorer (COBE) satellite image of

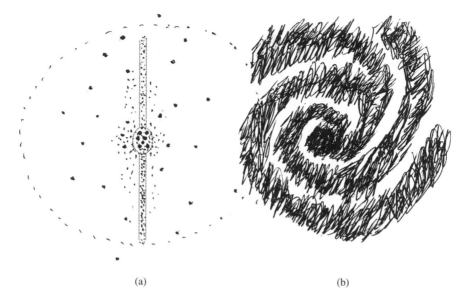

(a)   (b)

Figure 15. Schematic representation of the Milky Way Galaxy. (a) Side view, with stars identified by the small black dots and clusters of stars by heavy black dots, the halo volume indicated by the dashed-line circle, and the disk and central bulge delineated by the faint solid lines. (b) A sketch of the Galaxy's likely shape as would be seen from above or below. Note the color inversion: stars and glowing gas and dust are black, empty space is white, just the opposite to reality.

the galaxy, found at the NASA's WMAP Web site (both Web sites are listed in the Bibliography).

Expansion and Hubble's Redshift/Distance Relation

Hubble made two significant findings in the 1920s, of which the extragalactic nature of the spiral nebulae is the lesser result, although it provides the underpinning of the other, more momentous one. This lesser result was based on just a few nebulae, and in the period 1925–1929 he measured the distances to a number of other (by current standards not-too-distant) galaxies, using both Cepheid variables and very bright stars as *standard candles*, the generic name given to distance indicators. He was also aware that many redshift studies on nearby nebulae had shown them to be receding—recall from the Doppler effect discussed in Chapter 3 that wavelengths are increased when the emitter is receding. Because nebular recession was puzzling to the astronomical community, Hubble wanted the redshifts for the somewhat more distant nebulae he was currently measuring, and he asked his Mount Wilson colleague Milton Humason to try to obtain them. This was not an easy task to accomplish, but Humason did so.

Now armed with both the recession speeds [from Equation (1)] and his own distance measurements for a chosen set of 24 galaxies—combined data that only he possessed—Hubble took two extraordinary steps. First, he decided to construct a plot of the red-shift-determined speeds *versus* the distances for the set of galaxies. (Although plotting one attribute against another was not unknown then (recall the H-R diagram), no one prior to Hubble had thought of comparing redshifts and distances.) Second, he drew a straight line through the data points, even though they showed considerable scatter (Figure 16). Readers who have had occasion to take a laboratory course in which data points needed plotting may recall the urge to use a straight line to represent the data points, whether or not such simplicity was justified. It is an almost natural first step to consider when plotting data, unless the data-taker/plotter has a reason to try a more complicated curve.

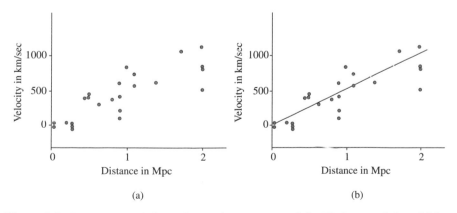

Figure 16. Recession of the galaxies/expansion of the Universe. (a) Hubble's 1929 plot of the recession speeds *versus* distances for the original set of 24 galaxies. (b) The straight-line fit to the data of (a). Adapted from Harrison (2000).

Edwin Hubble, however, might well have had a reason for trying the straight line, but before identifying it, I want to examine the pair of electrifying conclusions that follow from the extraordinary steps he took. The first was that the Universe is expanding: the more distant galaxies are receding from the solar system. This conclusion was electrifying because it not only contradicted the then conventional wisdom concerning the static Universe but also led to a major paradigm shift. However, it was probably not too surprising much less astonishing to Hubble himself: he had been aware of the theoretical possibility that the Universe could be expanding. (That the expansion is of space itself was a later interpretation, one I shall examine in Chapter 6.)

Hubble's second conclusion, based on his chosen set of galaxies, was that the expansion is occurring in a truly remarkable way: the speeds of these receding galaxies are proportional to their distances from the earth. This conclusion follows directly from his fitting a straight line to the scattered data points on the speed/distance plot. It is a precursor to the theoretical concept that the expansion is a natural consequence of certain characteristics that our Universe enjoys and is also related to the so-called de Sitter effect, alluded to above and described in Chapter 6. This is the effect that might have motivated him to draw the straight line through the data.

Expressed in mathematical terms, the proportionality Hubble inferred between recession speed V and distance D reads $V \propto D$, where $\propto$ again means "is proportional to." Normally, an expression of proportionality is converted to an equality by inserting a "constant of proportionality" into it. However, in the case of Hubble's original plot, the speeds he used were obtained from the approximate relation (1), and therefore, the expression $V \propto D$ itself is an approximation. This means that on inserting the constant of proportionality, one gets not an exact but an *approximate* equality. Bearing this in mind, and denoting the constant of proportionality by the symbol h_0, the approximate equality becomes

$$V \cong h_0 \times D. \tag{2}$$

Relation (2) is the mathematical equivalent of the straight line Hubble drew through the data points. In Chapter 6, you will encounter the *exact* theoretical counterpart to this relation. Known as Hubble's law, it involves a quantity denoted H_0, which is an analogue to h_0. I have introduced H_0 at this point to reemphasize the approximate, empirical nature of h_0. Each of these quantities is referred to as the Hubble constant, but until the theoretical basis for the exact relation was known, h_0 alone enjoyed that name.

In addition to relating the speed of expansion to the distance away, the Hubble constant is an ingredient in the quantity known as the critical density, which, in a uniformly expanding universe, determines whether or not gravity will reverse the expansion. Also, the large-scale curvature of the Universe, or equivalently, its large-scale geometry, depends in part on the Hubble constant, whose inverse provides an estimate of the age of the Universe. The attempts to determine the value of this constant lasted many decades, and it was not until the 1990s that a relatively accurate number was obtained, first using supernovas and then the CMB. These efforts are described in Chapter 7.

The Hubble constant can be stated as the ratio of a speed to a distance, or equivalently as an inverse time. The speed/distance ratio is obtained by dividing both sides of (2) by the quantity D, which gives $h_0 \cong V/D$. In this form it is expressed as so many km/sec per Mpc. This ratio of units can be reduced to an inverse time by using the facts that the dimension of speed is length per time, whereas that of distance is length. The dimension of V/D then becomes (length/time)/length. On canceling the common factor length, the dimension is just inverse time (unity over time). Equivalently, the dimension of both $1/h_0$ and $1/H_0$ is time.

Hubble, of course, was the first person to evaluate the constant now named for him. He obtained $h_0 \cong 530\,\mathrm{km/sec}$ per Mpc, or $1/h_0 \cong 1.8$ billion years. To assign a meaning to this number, it is helpful to recall the speed/distance analogy on page 78 concerning the lifetime of the sun. There I noted that the time required to travel $30\,\mathrm{km}$ at a constant speed of $60\,\mathrm{km/hr}$ is 30 minutes. This value is the result of dividing the distance traveled by the speed. The same ratio of distance divided by speed can be obtained from the relation $h_0 \cong V/D$ in the preceding paragraph by inverting each side. It leads to $1/h_0 \cong D/V$, which shows that the inverse of the Hubble constant is the ratio of the galactic distance to its speed. Let us now assume, exactly as in the driving example, that the expansion of the Universe occurs at a constant speed starting from an initial instant (the Big Bang?!). Under this assumption, $1/h_0$ becomes the time elapsed until the moment of measurement, in other words, the *lifetime* of the Universe.[7] From Hubble's analysis, that number is approximately 1.8 billion years.

This was the first scientific estimate of the age of the Universe, and it turned out to be less than the age of the earth! Something was clearly wrong. The mystery involved a stellar property that was only discovered in 1942 by the German astronomer Wilhem Baade, who had sought refuge in the United States from the Nazis prior to the Second World War. His status as a legal alien had prevented him from joining the U.S. armed forces, an event that allowed him almost unlimited observing time on the Mount Wilson telescope during the war and led to his discoveries.

The initial step in unraveling the mystery of an h_0 that produced a too-short-lived Universe was Baade's 1942 realization that there are two types of stars, known as Populations I and II. The former are young stars, which develop in the dust clouds located in the spiral arms of galaxies; the latter are old stars that are found in globular clusters, elliptical galaxies, and the dust-less regions of spiral galaxies. Baade's ultimate decoding of the h_0 mystery involved his second discovery, made in 1952 while working at the new 200-inch telescope at Mount Palomar in California. There he found not only that Populations I and II Cepheids had different period–luminosity (P-L) relations, but that Hubble, while observing type Is, had used the P-L relation for type IIs to analyze his results.

The immediate result of Baade's finding was an increase in the Cepheid distance scale by a factor of two, followed soon after by the American astronomer Allan Sandage further revising the distance scale. His revision reduced Hubble's value of h_0 by a factor of three, to about 180 km/sec per Mpc, thereby increasing $1/h_0$ to approximately 6 billion years, an estimated age for the Universe greater than that of the solar system.

Baade's and Sandage's revisions of some of the cosmic distance scales coupled with the changes in h_0 were the impetus for a variety of astronomical activities. These included checking older distance-measuring techniques and searching for newer ones, plus remeasuring the value of h_0 (later H_0). Sandage, who played a seminal role in this work, obtained the value (55 ± 10) km/sec per Mpc for the Hubble constant. For decades he insisted on the correctness of this value, despite the claims of other investigators that it was greater by as much as a factor of two. This led to a long-running controversy, which is summarized by Webb (1999). Readers interested in scornful scientists and some of their comments may find his account amusing.

The problem of determining a consistent and presumably correct value of the Hubble constant will be considered in Chapters 6 and 7, after I discuss the meaning of H_0. To obtain a correct value, however, requires that one has at hand a set of reliable distance indicators, an observational feature I turn to next.

Some Other Steps on the Cosmic Distance Ladder

Astronomers and cosmologists have long had a very deep seated, even a passionate, interest in finding accurate distance indicators. The motive of greatest interest to this chapter is validating the expansion of the Universe via the speed/distance relation. A wide-ranging summary of the main methods underlying construction of the cosmic distance ladder is given by Webb (1999), who emphasizes the attendant uncertainties, even for relatively near distances. I will touch on only a few them; if you want further information and feel ready to cross swords with some technical matters, you should find Webb's book a fascinating romp in the field of heavenly distances.

The first of the standard candles to be recalibrated were the Cepheids. The recalibration was (and is) carried out by comparing Cepheid-based distances with those from other procedures, assuming, of course, that the other methods yield reliable numbers. A few such comparisons are displayed in Table 9, where Candle 1 refers to RR Lyrae stars (a type of variable star), and Candle 2 estimates come from analysis of the so-called tip of the red giant branch of the Hertzsprung–Russell diagram [see Webb (1999)].

Table 9. Checking Cepheid Distances

Object	Cepheid	Candle 1	Candle 2
M31	733	740	773
M33	843	815	871
LMC	50	45	48

From Webb (1999). All distances are in kpc.

The overall variation in distances ranges from about 4% to 10%. The uncertainties are not prohibitively large, and the values are consistent with one another. Nonetheless, even for a galaxy as relatively "close" as LMC (the Large Magellanic Cloud), the distance is still somewhat ambiguous, in that the uncertainty is not at the 1% level. And that is precisely the point of the table: it is a reminder that no distance method is error-free, so that the higher you go on the cosmic distance ladder, the more rickety it becomes.

The distance range covered by Cepheids is roughly 15 kpc to 10 Mpc. Among the standard candles that reach beyond this upper limit are *type Ia supernovas* and *gravitational lensing*. Since their reach is greater than ones on rungs below them, they have played major roles in current research. However, the expansion of the Universe becomes significant for the extremely large distances measured by them, with the result that the usual notion of distance is no longer valid, a feature I examine in Chapter 7.

Type Ia supernovas are believed to be explosions of white dwarf stars whose mostly carbon mass is initially close to the Chandrasekhar limit of $1.4 M_{Sun}$. The explosion occurs in binary systems where the dwarf is the companion of an ordinary star. Just as with a black hole or neutron star in a binary star system, the white dwarf's intense gravitational field sucks matter off the companion star, matter that slowly accretes onto the dwarf's surface. As this happens, a series of nuclear reactions on its surface turns the dwarf into an extremely carbon-rich object. Meanwhile, its mass gradually increases, eventually exceeding the Chandrasekhar limit. As a result, the dwarf's temperature becomes high enough ($\cong 10^9 K$) that the carbon nuclei fuse together and, in a not entirely understood process, the white dwarf disintegrates in a cataclysmic explosion.

Although the total energy released in the explosion does not differ significantly from that of a type II supernova, almost all of it appears as electromagnetic radiation, in contrast with the type II case, where much of the energy output is in the form of neutrinos. The maximum luminosity of type Ia supernovas is about 5×10^9 as great as that of the sun—in other words, types Ia can also be as bright as a galaxy. The enormous amount of energy released causes them to burn out fairly rapidly. Most importantly, almost all type Ia supernovas have the same maximum luminosity, and they have therefore become excellent long-distance standard candles. However, there is a minor shortcoming to this rosy picture: type Ia supernovas have been calibrated by comparing them with Cepheid variables located in the same galaxy. As a result, the uncertainties in the distances inferred from type Ia supernovas cannot be less than those from Cepheids.

Although their value as a standard candle had been recognized, the relatively small number of type Ia observed at one time had been a drawback to their use. That situation changed with the advent in the 1990s of two international collaborations, the Supernova Cosmology Project, headed by the American physicist Saul Permutter, and the High-Z Supernova Search Team,[b] led by the American astronomer Adam Reiss. These collaborations found and were able to calibrate many type Ia supernovas, a major result of which was their discovery of the acceleration of the Universe, discussed in Chapter 7.

Gravitational lensing is another extremely long-distance candle that allows astronomers and cosmologists to look back billions of years in time. It enjoys the additional and important feature of providing information on the mass of the emitter whose

---

[b] Z in "High-Z" is the standard symbol for the redshift parameter introduced on page 44, defined by $z = (\lambda_{\text{obs}} - \lambda_{\text{emit}})/\lambda_{\text{emit}}$, the quantity on the left-hand side of Equation (1). It is one of the most important links in cosmology between theory and observation. In terms of it, Equation (1) reads $z \cong V/c$, while in many Hubble plots, the V on the vertical axis is replaced by z or zc, and the graph is one of redshift *versus* distance.

radiation is being affected. Gravitational lensing arises from the *warping*—or *distortion*—of space by the presence of large amounts of matter. Such warping, which I describe in the next chapter, is a phenomenon exclusive to Einsteinian gravity. Take a look at Figure 19 if you can't wait to see an illustration of such warping.

As a point of orientation, I will begin with a few comments about lenses. They are familiar objects in everyday life, prime examples being those in the eyes of living creatures, in eyeglasses, and in sunglasses. They also are essential ingredients in optical devices, for instance telescopes, binoculars, and microscopes. For persons, the basic function of a lens or combination of them is to focus light onto the retina of the eye. Light is also focused by mirrors, in particular spherical or parabolic ones. Figure 17 sketches the basic process by which such focusing occurs, the upper portion illustrating it for a spherical mirror, the lower for a *converging* lens. In each case, parallel rays of light are impinging on each of these objects, with the direction of the rays indicated by the arrows.

For the case of the mirror, all the light is *reflected* by the spherical surface and focused at the focal point *F*, after which each

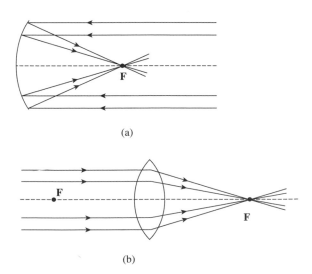

(a)

(b)

Figure 17. Focusing of parallel rays of light. (a) Portion of a large-radius spherical mirror that focuses the light at the focal point **F**. (b) A converging lens, which *refracts* or bends the light to focus it at the far-side focal point **F**.

ray continues on. *Reflecting telescopes* have at their base a mirror that focuses the light, which is then directed to a viewing or recording eyepiece (or set of them). In the case of a converging lens,[c] the light rays are *refracted* (or bent) as they pass through the lens, meet at the focal point, and then continue on. Because of the left–right symmetry of the converging lens, there is a second focal point to its left.

Both the mirror and the converging lens change the direction of the paths followed by the rays of light. Although it is a far-fetched use of the terminology, one might say that because of this directional change, these optical objects could be considered to act *as if* they distorted or warped the space in which the light rays are propagating, causing the change of direction. This only means that if they were absent, the light would continue to travel in a straight line. Of course, the space is not deformed, this is just an "as if" scenario.

In contrast to this *as if*, the general theory of relativity states that otherwise empty space *is* distorted by the presence of a mass in it. One effect of this distortion is to *deflect the paths of both material objects and radiation*, thus giving them the *appearance* of having been acted on by the force of gravity, exactly as in the classical theory of Newton, to which the general theory of relativity reduces when the masses are not too large.

A graphical illustration of the distortion and its effects makes use of what Harrison (2000) has labeled the expanding rubber sheet universe (ERSU). Imagine a stretched, flat sheet of rubber, whose two-dimensional surface is to be thought of as representing ordinary three-dimensional space. Because the sheet is made of rubber, it is stretchable and so functions as a deformable universe. As long as no masses are present, the sheet remains flat, but insertion of a mass deforms it, in much the same way that placing a heavy weight on a mattress depresses it. Figure 18 shows the undistorted flat sheet in the absence of any mass.

In Figure 19, a very large mass M has been placed into this previously empty universe, the result of which is the big

---

[c]In diverging lenses, which do not enter our discussion, the rays of light are spread out rather than being focused. The lenses in eyeglasses worn by very near-sighted persons are of this type.

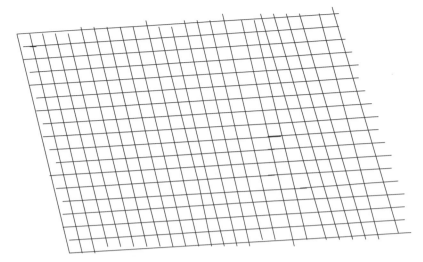

Figure 18. Sketch of a flat portion of space in the expanding rubber sheet universe of Harrison (2000). The cross-hatched area represents undistorted (i.e., flat) space in the absence of mass.

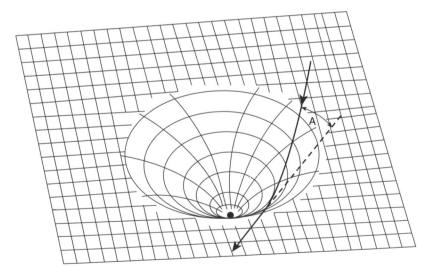

Figure 19. Distortion or warping of the same portion of space shown in Figure 18 by the presence of a heavy mass, indicated by the large black dot at the bottom of the depression. The solid line with two arrowheads is the deflected path of a light ray passing near the distorted spatial region, and the dashed line indicates the direction that an observer would "measure" the deflected ray to have come from.

depression seen in the figure. The depression represents the warping due to the mass, itself shown as the heavy black dot at the bottom of the distorted sheet. In terms of dimensions, M causes the two-dimensional ERSU to become three-dimensional— and this is why a flat ERSU was used: there is no simple way to represent the distortion if a three-dimensional space had been chosen to represent the initially massless universe.

Also exhibited in Figure 19 is a light ray (a beam of photons), assumed to be emitted by a distant source such as a star. The ray's trajectory is the heavy solid line with two arrows in it. When the photons are far from the mass, their path is shown as the straight line with the arrow in the top part of the trajectory. Once the photons are close enough to M to respond to the spatial distortion, they do so by changing their trajectory to the curved part of the solid line. Finally, after the photons have gotten far enough away from the mass that the distortion is negligible, they again follow a straight-line path, indicated by the portion of the trajectory containing the lower arrow.

Suppose that a telescope were to detect the beam of photons after they traversed the distorted region of space. An observer would infer that the photons were emitted not from the original source but from an *apparent* one in the direction pointed to by the final straight-line path, indicated in the figure by the dashed line. The difference between the actual direction of emission and the observed direction is the angle A in the figure. This deflection of the path of a light ray passing near a localized mass is an example of *gravitational lensing*, whose signature here is the nonzero angle A, which theory states is proportional to the mass M.

Deflection of light by a mass—for instance by the sun —is also a feature of Newtonian gravity, to which the term *gravitational lensing* is equally applicable. In 1801, the German astronomer Johann von Soldner became the first person to use Newtonian gravity to calculate A for the case of a light ray just grazing the surface of the sun; he found the value 0.875″. Unaware of this result, Albert Einstein recalculated it in 1911, during the 8-year period when he was developing the general theory of relativity, whose gravitational effects—as already emphasized—arise from the distortion of space by the presence of mass, as illustrated in Figure 18. In 1915, Einstein perfected the final equations of the theory; among other applications, he used them to calculate the deflection of light as it grazed the sun's surface. The answer he got was $A = 1.75''$, twice the Newtonian value.

After Einstein published his result, the natural question was which theory of gravity is correct, Newton's or Einstein's. The First World War prevented definitive tests from being carried out, but in 1919 Arthur Eddington organized two expeditions to measure the angular position of a star: first during a solar eclipse (so that the starlight grazing the sun's surface could actually be seen), and then 6 months later for comparison. The photographs that were taken clearly established the correctness of Einstein's theory over that of Newton, but instead of this landmark result being typically confined to the scientific community, it became a worldwide news item. The international excitement was enormous, and Einstein, whose previous work had already made him a celebrity in the scientific world, turned into a far greater one in the nonscientific community.[8] Newspaper phrases such as "a revolution in science", "a new theory of the universe", "Newtonian ideas overthrown", etc., catapulted him almost overnight into the best-known person in the world, the type of event that was a precursor to the tabloidish, celebrity-obsessed culture so dominant in our own times.

By successfully passing the deflection-of-light and other tests (some of which I will review in the next chapter), the general theory of relativity has become the paradigmatic theory of gravity, albeit one in which quantum concepts do not enter. The phenomenon of gravitational lensing was eventually identified as a possible cosmic distance indicator in which a galaxy or cluster of galaxies, with mass much greater than M_{Sun}, could focus light otherwise too faint to be seen. Unlike the single stellar image seen in the 1919 photos, gravitational lensing has often produced multiple images of the source, as well as arcs and rings. Examples can be found at the Hubble Space Telescope Web site, listed in the Bibliography.

Among the classes of object to which lensing has been applied are *quasars*, an acronym for the phrase *quasi-stellar radio source*, coined in 1964 by the Chinese-American astrophysicist Hong Yee Chiu. Quasars, only relatively few of which actually emit much of their enormous energy at radio frequencies, were a puzzle when

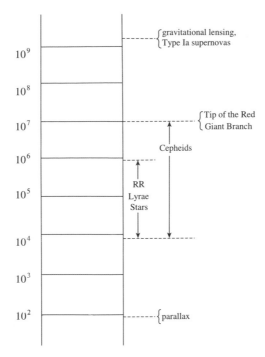

Figure 20. Some rungs on the cosmic distance ladder, where the vertical scale is given in parsecs and only those methods cited in the text are shown. The approximate distance ranges for RR Lyrae stars and for Cepheid variables are indicated by the arrows, and the rough upper limits to the distances yielded by parallax, by the Tip of the Red Giant Branch method, by gravitational lensing, and by type Ia supernovas are identified by the top two and the bottom dotted lines. Distances from Harrison (2000) and Webb (1999).

first observed, essentially because their spectra were unidentifiable. The puzzle was solved by the Dutch-American astronomer Maarten Schmidt in 1963, who realized that the spectrum of the quasar denoted 3C273 made complete sense if it were redshifted by 16% ($z = 0.16$). The spectra of other quasars have been similarly interpreted as arising from similar "largish" redshifts. Such redshifts are large only in comparison with the very small ones obtained by Hubble, which did not exceed 0.004, whereas ones greater than 6 have been found. These correspond to objects that are exceedingly far away yet are very compact emitters of energy. They are now believed to be highly luminous galaxies whose turbulent centers contain supermassive black holes.

Depending on the definition of distance used in an expanding universe context, the distance to some lensed galaxies is greater than 3000 Mpc. Though very large, this distance does not reach the edge of the observable Universe, which I estimate in Chapter 7 is about 50% farther away.

I have now identified a variety of steps on the cosmic distance ladder, and even though the term *ladder* is metaphorical, Figure 20 shows a pictorial representation of one. The four methods mentioned in this chapter, plus the ones from Chapters 2 and 3, are indicated on the right side of the figure. The approximate ranges yielded by use of RR Lyrae stars and the period–luminosity relation for Cepheid variables are specified by the vertical arrows, while the rough upper limits from the other four methods are indicated by the dotted lines next to their names. The vertical scale is in parsecs.

The procedures identified in the figure are members of a larger set, some details of which can be found in Harrison (2000) and Webb (1999). A comprehensive compilation of distance-measurement methods is given in Figure 11.6 of Webb (1999).

6. Homogeneous, Isotropic Universes

Hubble's discovery that the Universe is expanding in such a way as to embody a speed–distance law raises a variety of questions. For example, What kind of Universe has such features, and what are its other properties? Observation alone cannot provide the desired answers; theoretical investigation is required as well. The theoretical developments that yielded answers to these and some of the questions raised in earlier chapters began in 1915 when Albert Einstein produced the final version of the general theory of relativity. The growth of this theory into the one that dominated research on cosmology for so long is the theme of this chapter. I will first review some of the relevant history and then describe the meaning and some implications of a homogeneous, isotropic universe.

Historical Background: Beginnings

By 1915, Einstein was extremely well respected in the scientific community for his prior research on the special theory of relativity and the photoelectric effect, among much else. In view of this and the interest his work aroused, it is highly unlikely—contrary to statements made in the past—that only a tiny number of scientists understood his theory of gravity when it first appeared in print. Both alone and in collaboration with a few others, he had published earlier versions of the theory; furthermore, he had corresponded with other scientists about it. (Recall that his articles had been read by Karl Schwarzschild, who used them to analyze the effects of gravity due to a point mass.)

In addition to the calculation of the deflection of light by a star, Einstein also used general relativity in 1915 to solve an outstanding puzzle in astronomy. The puzzle concerned the *precession* of the perihelion of the planet Mercury (the perihelion is the distance of closest approach of a planet to the sun). Because of perturbing effects due to the other planets, the major axis of the orbit of any one of them will, over time, very slowly change its orientation in space, a movement known as precession (see Figure 4). Mercury, with its large eccentricity (Table 2), was the planet whose precession was easiest to measure. This had been done, and a plot of its perihelion *versus* time had shown an advance due to the precession. Calculations based on Newtonian gravity, which described the advance qualitatively, gave the wrong answer! It failed to account for an extra advance of 43" of arc per century—a small number but well outside the error of the measurement. Einstein's theory, on the other hand, gave the right answer, including the extra 43" of arc per century. This stunning result helped establish general relativity in the scientific community as a possible new paradigm for gravity. The successful measurement in 1919 of the deflection of light by the sun then became a key factor in establishing this possibility as very likely.

This likelihood has become a certainty, and the general theory of relativity is now the paradigmatic theory of gravity.[1] In producing gravitational effects through the warping of space (and also of time) by mass and/or energy, it is a theory that contains unfamiliar elements. One of them is the generalization of flat, Euclidean space—the "ordinary" geometry we are accustomed to in our own locales—to other types, ones wherein there is an inherent curvature to space. The type of curvature that occurs is influenced, among other factors, by the presence of mass. Both Figure 18 and black holes exemplify this particular influence. The type of curvature, as well as its possible absence, depends on the value of a parameter that enters the equations of general relativity, and I shall postpone consideration of other geometries as well as curvature, particularly that of our Universe, until the next chapter.

In 1917, two years after the deflection and perihelion calculations, Einstein obtained his first full solution to the equations of general relativity. It described a universe that could either expand or contract. In order to conform to the prevailing belief that our Universe is static, Einstein made a "slight modification" to his equations, adding to them a term containing what is now known as the *cosmological constant*. This new term acts like a repulsion, and he adjusted its value so as to exactly counter the attraction due to gravity, thereby achieving a static universe. Hubbell's later discovery of galactic recession means that the Universe is *not* static. This implies that the new term is superfluous, an implication Einstein accepted: he died believing that its introduction was "the greatest blunder of my life."[a] Rather than a blunder, however, the cosmological constant has become one of the most significant constants in cosmology. As you will find in the next chapter, it can account for the observed acceleration of the Universe, and its measurement and interpretation has been the subject of much recent research.

In 1917, the Dutch astronomer Willem de Sitter published a second solution to the equations of general relativity. This one, which also embodies the effects of a cosmological constant, describes a universe that expands at a constant rate but in such a way that it always looks the same. The unvarying aspect means that de Sitter's theoretical universe is *stationary*, rather than static; it is also infinitely long-lived. Thirty-odd years later, it became the foundation for the *steady state theory* of the Universe, described in the next section.

de Sitter realized that the spectra of distant stars would be redshifted in the universe generated by his solution, although he also stated that this result was spurious! As a corollary to what was not a spurious result, the German mathematician Hermann Weyl pointed out in 1923 that these redshifts would increase as the distance to their source increased, thereby anticipating Hubble's finding. [Bernstein and Feinberg (1986) point out that this conclusion is actually implicit in de Sitter's paper.] However, Weyl, just like all the others who wrote about expansion and redshifts prior to Hubble's discovery, had no observational basis for claiming that the idea applied to our own Universe. It is not known whether Hubble was aware of Weyl's result, as Hubble does not cite Weyl's article—but then again he doesn't cite de Sitter's papers either, even though he *refers* to the de Sitter effect, as I noted earlier. It is this awareness of de Sitter's work—and possi-

---

[a]As quoted by the Russian-American physicist George Gamow in his autobiography *My World Line* [Gamow (1970)]. Gamow became one of the champions of the Big Bang origin of the Universe, as I recount later in this chapter.

bly that of Weyl—that underlies the comment on page 108 that Hubble might have had a definite reason for trying a straight-line fit to the redshift/distance data.

The next significant theoretical step was taken in 1922 by the Russian atmospheric physicist Aleksander Friedmann, who generalized the solutions to the equations of general relativity obtained by Einstein and by de Sitter. Among Friedmann's solutions is one corresponding to an expanding universe that has a definite origin in time: it is a Big Bang type of universe! This was not a concept for which the scientific community was ready, because as Friedmann acknowledged, there was then insufficient evidence to decide if *any* of his theoretical universes might correspond to our own. There is a wry aspect to Friedmann's work: in one of those kinds of critical lapse already noted, Albert Einstein mistakenly faulted Friedmann's research in a short note published in 1922. Less than a year later, Einstein retracted the criticism.

Historical Background: The Modern Era

The foregoing milestone developments arguably belong to the premodern era of cosmological research, whereas one might reasonably claim that the modern era begins with the 1929 paper of the American mathematical physicist Howard Robertson. This claim rests on the fact that Robertson was the first person to analyze *systematically* the effect of certain symmetries on the equations of general relativity. As others had before him, but not systematically, Robertson employed the concepts of *spatial homogeneity and isotropy*, terms defined in the next section.[b]

I cannot stress too strongly the significance of symmetry concepts in physics and in cosmology in particular. For example, Hubble's *law*, which is the name given to the *exact* speed/distance

---

[b]While Einstein, de Sitter, and Friedmann had used homogeneity and isotropy to simplify and solve the equations of general relativity, Robertson was the first to undertake a systematic investigation of the role of symmetries in cosmological research, so he is credited with inaugurating the modern era.

relation $V = H_0 \times D$, is a direct consequence of the fact that on the very largest scales, and to an accuracy of about 1 part in 10^5, our Universe is both homogeneous and isotropic. Later in this chapter, I will use these two characteristics to demonstrate the existence of the so-called scale factor of the Universe and to derive the just-stated Hubble law, as well to show that the Hubble constant H_0 is related to the scale factor and its rate of change in time. Heady stuff, and highly significant.

One might also claim that the modern era in cosmology, opened by Robertson's 1929 paper, was—at least in retrospect—firmly established by the 1931 article of a Belgian cleric, the Abbé Georges Lemaitre. He, too, used the concept of spatial homogeneity to simplify the equations of general relativity, expressing them in the form found in some modern textbooks. Most significantly, he used the universal scale factor to *derive* the Hubble law and raised but did not answer the question of why the Universe is expanding. The cosmological constant appears in this article, which describes a universe that expands forever. In later work, Lemaitre recognized that an expanding universe would have started explosively at some much earlier point in time in a highly condensed and hot state, which he variously called the "primeval atom" or the "cosmic egg." Lemaitre's use of the word "egg" as the origin of the Universe followed a long tradition of this usage in various cosmologies, as pointed out by Bernstein and Feinberg (1986) and by Teresi (2002).

Lemaitre's assertion that our Universe originated as a primeval atom identifies him as the "father" of the Big Bang origin of the Universe, even though Friedmann's equations described an expanding universe with a definite origin in time. The difference in this regard is one of timing: Friedmann had no observational evidence for the expansion, whereas Lemaitre, who re-derived Friedmann's results and obtained others, knew of both the expansion and the speed/distance relation.

Although Friedmann's death in 1927 cut short the possibility of his making further significant contributions to the study of cosmology, he strongly influenced someone who did. This was the Russian-American physicist George Gamow, who, along with the Americans Ralph Alpher (his former Ph.D. student) and Robert Herman, established the field of early-Universe studies. From 1946 to 1953, Gamow, alone and with his collaborators, published a series of papers that examined the behavior of an expanding, Friedmann–Lemaitre type universe that at very early times contained a high-temperature mix of neutrons, protons, neutrinos, and photons (a mix he called the *ylem*). Using what was then known quantitatively about the necessary nuclear physics—which was quite a lot—Gamow incorrectly concluded that his analyses were able to describe the primordial formation of the *entire* set of the then-known chemical elements as well as the separation of the galaxies into individual structures. Although this was overreaching, the analyses and predictions of Alpher and Herman were "right on the money," though no one knew it at that time.

In 1948 and 1949, Alpher and Herman, building on Gamow's analyses, predicted that photons from the early Universe should exist in our era as blackbody radiation at the very low temperature of approximately 5 K. (At *early* times in a Big Bang universe, of course, the photons are at a very high temperature, but as such a universe expands, it cools and so does the radiation, with the result that the photons are described by a blackbody spectrum corresponding to lower and lower temperatures.) Gamow made similar predictions concerning the photons: in 1950, he stated the current temperature to be 3 K; in 1953, he estimated it to be at the slightly higher value of 7 K. As noted in Chapter 1, such predictions imply a direct test of the assumption that the Universe evolved from a Big Bang origin. In particular—and in hindsight—the 1948 prediction of Alpher and Herman ought to have been subjected to the test, since a competing cosmology had been put forward.

The rival cosmology is the steady state model (of continuous creation) proposed in 1948 by the British astronomers Herman Bondi and Thomas Gold (identified in Chapter 1). They envisaged an expanding universe that obeyed the *perfect cosmological principal*, which states that the Universe is the same at all places and for all times (it was a stationary, de Sitter–type universe, as noted earlier in this section). Because their universe was infinitely old, it neatly circumvented the then problem that a Big Bang Universe with a Hubble constant of 530 km/sec per Mpc would be younger than the solar system. (Recall that Walter Baade solved this particular problem only in 1952, 4 years after the work of Bondi and Gold.)

Although stellar and galactic evolution is allowed in the steady state model, the fact that it describes an expanding universe means that eventually the average density of matter in it will decrease. Because this would violate the perfect cosmological principle, Bondi and Gold proposed that matter is spontaneously and continuously created in their model, at a rate just sufficient to compensate for the decrease of average density. The required rate was roughly one proton (H atom) per cubic meter every billion years! Although this number is too small to be measured, the model has consequences that can be tested experimentally, although it was not done.

The steady state scenario attracted adherents, one of them being Fred Hoyle, who became a passionate advocate and in 1948 proposed a mechanism whereby matter could be continuously created within the structure of general relativity. Hoyle was publicly dismissive of the Friedmann and Lemaitre cosmologies, with their explosive origins at a finite time in the past. As already remarked, he denigrated the idea of an explosive origin by referring to it as nothing more than a "Big Bang." This once-pejorative name has, of course, stuck; in contrast, the eventual detection of the low-temperature blackbody radiation and of quasars from the early Universe led to the demise of the steady state model; Hoyle later published a retraction.

Although the Big Bang scenario has triumphed, in the context of the present historical review the key words in the preceding sentence are "eventual detection." The prediction of the pervasive, low-temperature blackbody radiation—the cosmic microwave background (the CMB)—was overlooked for more than 10 years by the scientific community in general (as I remarked in the first chapter, there was then no cosmology community to speak of). Not only did this prediction fall into limbo,[2] but so did the detailed calculations on the *nucleosynthesis* of deuterium, helium, and lithium nuclei in a hot, explosive early universe that were published by Alpher and Herman in the late 1940s and by Alpher, Follin, and Herman in 1953 (described in Chapter 8 and Appendix B).

No experimental searches for the CMB were carried out in those 10 years, and the controversy between the two competing theories on the nature of the Universe remained largely confined to the participants and their associates. Things might have been different had cosmology been recognized as a worthwhile topic of research, but the overwhelming majority of physicists and astronomers were not yet ready to acknowledge this possibility. Furthermore, the experimenters and the general relativists tended not to communicate. The consequence was a scientific community either ignorant of the prediction/controversy or uninterested in it.

The discovery of the CMB dramatically changed this situation, as noted in Chapter 1. I shall only give a brief summary of this change, which occurred in a serendipitous manner and has been the subject of various books and articles. If you wish to learn more about the curious details of this story, I recommend the discussions in Bernstein (1993), Bernstein and Feinberg (1986), Kolb (1996), and Weinberg (1977)—among others—especially for the roles played by other than the major participants. See also Lemonick (2005), who has written a history of the WMAP enterprise, and who recounts prior, but unknowing, observations of the CMB. (In contrast with the Penzias and Wilson discovery, the significance of these earlier observations was not understood.)

In the period 1964–1965, two American radio astronomers employed by Bell Telephone Laboratories, Arno Penzias and Robert Wilson, had been observing anomalous signals in their radio-frequency detector, signals whose origin was unknown, whose existence was problematic—they even removed pigeon droppings from their detector in a vain effort to eliminate what might have been a source of their anomalous signals!—and whose interpretation was a puzzle.

The measured wavelength of the radiation was 7.35 cm (in the microwave range), the temperature was approximately 3.5 K, and the radiation was isotropic (that is, unvarying with respect to changes in the direction of observation). These observations were consistent with the radiation predicted by Gamow, Alpher, and Herman, but, like almost every scientist at the time, Penzias and Wilson were unaware of the prediction.

They learned of it through eventual contact with a research group in the Princeton University physics department. This group, led by the American physicist Robert Dicke, was tooling up—both experimentally and theoretically—to look for a type of radiation similar to that predicted by the Big Bang theorists, although the radiation source Dicke had in mind was based on a different cosmology. Once the Bell Labs and Princeton researchers got together, the likely interpretation of the Penzias–Wilson results as relic radiation from a Big Bang explosive origin to the Universe was seized on, the scientific literature was searched, and two articles were written and submitted for publication. That of the Princeton group dealt with the interpretation; the discovery reported by Penzias and Wilson won them the Nobel Prize in physics.

Once the physics community realized that Penzias and Wilson's discovery might be relic radiation left over from a Big Bang explosion in the very early Universe, major efforts were undertaken to measure the CMB over its full spectrum of wavelengths/frequencies. These culminated with the 1989 launching of the Cosmic Background Explorer (the COBE) satellite. Its 1992 results provided the desired information (the measurements lasted into the mid-1990s). The CMB is indeed blackbody, with a peak in the microwave region and a temperature of 2.725 K. It is the chilled radiation from a Big Bang event in the very early Universe. And, as already indicated, the CMB is isotropic (uniform) over the whole sky to 1 part in 10^5, at which level *anisotropies*, or deviations from uniformity, arise. Since these anisotropies yield extremely important information about the Universe, their investigation has been the subject of both observational and theoretical studies in recent years.

The discovery of the CMB, and its later measurement over the full wavelength range, are events of the utmost significance to modern cosmology. Among the many ramifications was the establishing of cosmology as a legitimate scientific enterprise comprising a community of researchers (who, naturally enough, labeled themselves cosmologists). The first graduate text on the subject appeared less than 10 years later [Weinberg (1972)]. In a more technical vein, both the uniformity in temperature of the CMB radiation over the entire sky and the deviations from this isotropy at the level of 1 part in 10^5 have had enormous consequences, ones I shall explore in the next section and succeeding chapters. The accuracy of these measurements is, of course, a triumph of modern-day technology. Readers interested in the experimental results can log on to almost any of the Web sites listed in the Bibliography; the general shape of the CMB curve, plotted in terms of frequency, is just that given by Figure 12b.

Once cosmology was established as an acknowledged scientific enterprise, it took little time for experimental and theoretical research in it to flourish. Research areas that were nascent prior to the detection of the CMB have been investigated in great detail. The construction of new observatories, both terrestrial and satellite based, has resulted in an almost continuous wave of new results, some totally unexpected, many generating much excitement, among which is the discovery that the Universe is accelerating, not simply expanding at a speed proportional to distance. (The discoveries of the expansion, the CMB, and the acceleration are undoubtedly the three most momentous cosmological observations of the 20th century).

On the theoretical side, there has been the vitally important realization that cosmology and elementary particle physics are closely linked, particularly in the area of the very early Universe, the topic of Chapter 9. However, the theory that is immediately relevant is that of the next section.

Homogeneity and Isotropy

To an excellent approximation, and on the largest scales—intergalactic rather than interstellar, the density of matter is essentially constant and is extremely small. For example, the density of luminous matter is approximately equal to 3.2×10^{-29} kg/m^3 (Table 7), or roughly one proton mass per 50 cubic meters. The total density of ordinary matter, which can shine but most of which does not, is about 4×10^{-28} kg/m^3 (from Table 12), or between 2 and 3 proton masses per 10 cubic meters. Even when dark matter is included, this number only increases to about 15 proton masses per 10 cubic meters. On average, the Universe is mostly empty space. In addition, the temperature of the cosmic microwave background radiation is also almost perfectly uniform. These two facts are the evidence underlying the proposition that the Universe is *homoge-*

neous and *isotropic*, terms I introduced earlier and am now ready to define.

Homogeneity means that one cannot distinguish one place from another along any direction. That is, in a homogeneous environment, there are no distinguishing features that allow one to identify particular points as one changes location: they all look the same. *Isotropy* means that one cannot distinguish one direction from another; analogously, there are no distinguishing features in an isotropic environment: the environment appears the same as one's orientation rotates about any particular point. Thus, if there are circumstances in which it is possible to identify either a location or a direction, then either homogeneity or isotropy, respectively, fails to be a valid concept for that situation. For instance, because of its streets and buildings, a city is neither homogeneous nor isotropic. Since temperature and density each decreases as one moves radially outward from its center, the sun is not homogeneous. However, considered from the special viewpoint of its center, the interior of the sun is believed to be isotropic: the decreases in temperature and density are each radially uniform. In other words, no matter which direction from the center one looks along, the inhomogeneous decrease is the same.

Terrestrial environments are rarely homogeneous and isotropic. An instance when both are valid is the following example, not entirely contrived. Imagine a fishing boat at sea and out of sight of land on a calm, completely overcast day. Because neither land nor the sun is visible, and the ocean's surface is assumed to be unvarying, there are no visual clues to location or direction. The two persons in the boat are thus in the midst of a homogeneous, isotropic environment.[3] Were the skies not overcast, the sun might function as a partial location (a latitude) indicator, and of course use of a modern position-finding mechanism like GPS negates the point of the example. However, even with a GPS, the persons on the boat are still in the dark, so to speak: it's the GPS that pierces the clouds and identifies the location, not the crew directly.

There are some extremely important technical consequences that follow from homogeneity and isotropy of the Universe, ones I shall soon explore. However, an essentially nontechnical aspect can be grasped immediately, and its implications are among the most thought-provoking of all. Recall that homogeneity means an inability to specify a particular location: all are equally unidentifiable. If all locations are equal in the preceding sense, then there cannot be a center to our Universe—or to any theoretically generated ones that are homogeneous. That is, no matter how hard you tried, no center could be found, since its existence would contradict the assumption of homogeneity.

The absence of a center to the Universe may raise a perplexing question in your mind; namely, if the expansion of the Universe is the result of an explosion from a highly condensed and hot state at very early times, how can there be an explosion or an expansion without a center? The answer to this question involves two separate concepts, neither of the garden-variety type.

First, the common meaning of an explosion as a violent event whose products expand into a volume in some region of space is not applicable. Instead, the expansion that followed the Big Bang *is of space itself*: prior to the expansion there was no space outside of the tiny volume the Universe then occupied. Ordinary explosions occur in existing space; but after the Big Bang, it is space that expands, creating itself as it moves outward. For a more technical statement of this phenomenon, see Chapter 7.

The second nonstandard concept is that an expansion—even one following an explosion—can occur *without a center*. This is perhaps most easily grasped through the balloon analogy. Consider the inflating of a spherical balloon with a uniform distribution of small coins glued to its surface. This surface, which is two-dimensional and has no center, represents the three-dimensional space of the Universe; the coins represent the galaxies. Imagine that there are sentient creatures living in the "galaxies" on the surface, but under the strict proviso that they are always unaware of the space interior or exterior to their universe, that is, to the third dimension. (We, on the other hand, *are* aware of it, but we are not part of the analogy.) Assume also that the balloon can be inflated by a device located inside it, a device that the inhabitants do not (and cannot) ever discern.

As the balloon inflates, its inhabitants will observe— somehow—that their two-dimensional universe is expanding *everywhere* and that all of the balloon-surface galaxies are moving apart from one another. The balloon-surface universe is homogeneous, isotropic, and has no central point, and in this sense is an almost perfect analogue—but only an analogue—for our three-dimensional Universe when it is considered on the largest distance scales. That is, like the balloon surface, our Universe is expanding without possessing a center. It is also worth noting that the balloon's surface is finite yet has no edge, characteristics that may be shared by the Universe.

Both the balloon's surface and the calm ocean on a cloudy day provide analogues to all or part of our homogeneous, isotropic Universe when it is considered on the Mpc scale. From the perspective of general relativity, our Universe and each theoretically generated, homogeneous and isotropic universe are to be analyzed by assuming that the galaxies (real or theoretical) are *structureless atoms* forming a very dilute fluid. In other words, on the gigantic scale being considered, the internal structure of any galaxy—its stars, globular clusters, dust, etc—is regarded as nondiscernable.

Structureless galaxies in such a universe therefore are analogues of atoms in chemistry, wherein atomic sizes are the only important distances, and atomic nuclei—their own internal structure, so to speak—are nondiscernable on the atomic scale and thus play no role in chemical processes. Correspondingly, and as a consequence of the Mpc scale, the relatively "tiny" galaxies not only function like atoms but also *do not increase in size when the Universe expands*; only space does. Such structureless, theoretical models of universes lack all the features of our Universe as we know it: totally absent are living creatures, stars, planets, interstellar dust, black holes, etc—objects, which, on the gigantic scale being considered, cannot be distinguished. They are analogous to the individual leaves on a tree seen from a great distance away. These models of universes are therefore mathematical frameworks into which quantities such as mass and radiation densities (e.g., galaxies and the CMB) may be inserted after the relevant evolutionary properties have been studied in detail.

This latter comment means that the topics of galactic formation, types, evolution and distribution, plus large-scale structure in the Universe will not be addressed within the current context: it requires a separate treatment. Interested readers may consult the books by Harrison, Peacock, Raine and Thomas, Rees,

Rich, and Silk listed in the Bibliography and various references cited therein.

Some Consequences of Homogeneity and Isotropy

From now on, the fundamental premise is that despite their size, structure, and energy output, galaxies are to be thought of as identical *points* in a homogeneous, isotropic environment undergoing uniform expansion. I will use this scenario, with its symmetry properties of homogeneity and isotropy, to demonstrate the existence of the universal scale factor previously mentioned. The next step will be the derivation of the *exact* Hubble law (the speed/distance relation), just by assuming homogeneity. As you will see, these two results are independent of the actual makeup of a universe; for instance, what kinds of and how much matter is in it. But, Hubble's law need not hold whenever a cosmological constant is both present and dominant, because in that case speed and distance are not linearly related.

Consider first three arbitrarily chosen galaxies, labeled A, B, and C, whose positions form the vertices of a triangle, as in Figure 21a. Their separations at the instant depicted in the figure are labeled D_1, D_2, and D_3. After some passage of time, the universe has expanded, and the separations have increased to the distances D'_1, D'_2, and D'_3, shown in Figure 21b.

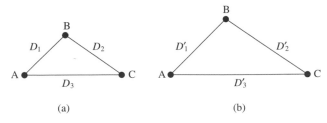

(a)   (b)

Figure 21. The consequence of isotropy on expansion. (a) Galaxies A, B, C, separated by distances D_1, D_2, and D_3 at some instant of time. (b) The same three galaxies at a later time, now separated by distances D_1', D_2', and D_3'. Because of the assumption of isotropy, the triangles in parts (a) and (b) are geometrically similar.

Because the universe under consideration is homogeneous and isotropic, the expansion must take place in the same way everywhere. Therefore each side of the original triangle expands in the same way, which means that geometrically, the two triangles in the figure are similar. Or, from the perspective of an observer on A, the uniformity of the expansion plus isotropy means that the percent increase of D_1 to D'_1 must be the same as the percent increase of D_3 to D'_3. Correspondingly, for an observer on B, the percent increase of D_2 to D'_2 must be the same as the percent increase of D_1 to D'_1.

All three distances therefore increase by the same factor, and since these are three arbitrary galaxies at arbitrary separations, it follows that *any* distance D will increase by this factor. It is denoted the *universal scale factor* (or the scale factor of the Universe, referred to previously), which I represent by the symbol R. Mathematically, R relates *any* expanded distance D' to its old value D by the equation

$$D' = R \times D. \tag{3}$$

This relation simply says that there exists the unique quantity R from which every expanded distance can be obtained by multiplication.

One of my emphases has been the use of appropriate distance units (e.g., km or ly). In order that each of D and D' is expressed in the same unit, the scale factor R would ordinarily and necessarily be expressed as a dimensionless number. In discussing homogeneous and isotropic universes, however, cosmologists usually flout the ordinary convention and remove the length unit from the old distance D, attaching it instead to R. It then follows from Equation (3) that D' obtains its length unit from R and not from D. Furthermore, since the numerical value of an expanded distance D' will vary in time, it is a second cosmological convention to allow R rather than D to vary in time as well. These conventions will be invoked a little later; now, however, I will show how to derive the exact speed/distance law.

The following derivation makes use of the assumptions of homogeneity and uniform expansion. Consider three arbitrary galaxies, again labeled A, B, C, lying on a straight line at identical

Figure 22. The consequence of homogeneity on expansion. (a) Galaxies A, B, C, lying on a straight line with common separations D, at some instant of time. (b) The same three galaxies at some later instant of time, still lying on a straight line but now with common separation D'.

pair-separations D, as in Figure 22a. Uniform expansion plus homogeneity means that at some later time each pair separation has increased by the same amount and become the distance D'. (Homogeneity *requires* that if the old separations are the same, the new ones are also.) The linear configuration at this later time is shown in Figure 22b.

Let us now ask, what are the new distances and speeds that an observer on A would measure? From A's perspective, B is seen to move from the separation D to D', while an observer on B would find that C has moved from the same initial separation D to the same final separation D'. Also, due to homogeneity and uniform expansion, all observers find every *adjacent* galaxy to be moving at the same speed, again denoted V. That is, were any galaxy to move relative to an immediate neighbor at a speed different than V when all distances between adjacent pairs of galaxies are the same, then a particular spatial location could be identified, thus violating homogeneity.

Hence, part of the answer to the question posed at the beginning of the preceding paragraph is that the observer on A would find that B is a distance D' away, receding at speed V, and that C is a distance $2D'$ away. Still to be determined is how fast C is receding from A. To obtain this result, the observer on A adds the speed with which C moves away from B to the speed that B itself is moving away from A. Each of these two speeds is V, and thus the observer on A concludes that C is receding at the speed $2V$.[4]

Summarizing these conclusions, we have found that B, which is a distance D' away from A, moves away from it at speed V, while C, a distance $2D'$ from A, recedes from it at a speed $2V$. That is, the speed with which each of the other two galaxies recedes from A is proportional to the distance of each of them from A. Furthermore, if there were a fourth galaxy on the same line at an initial distance D to the right of C, the preceding conclusion would again have been reached. Since the distances D and D' are arbitrary, then just as in the case of the universal scale factor, this conclusion can be generalized and expressed as the result: *in a uniformly expanding, homogeneous universe, speed is proportional to distance.*[5] Denoting the constant of proportionality by H, the speed by V, and the distance by D, the latter proportionality becomes the equation

$$V = H \times D, \tag{4}$$

which is the *exact form of Hubble's law. H* is known as the *Hubble parameter.*

Equation (4) is an exact speed/distance relation, one that results from homogeneity in a uniformly expanding universe; it has *nothing* to do with redshifts, in contrast with the approximate relation (2) that Hubble deduced. Equation (4) holds for all times under the conditions just stated. The Hubble parameter H varies in time—I will relate it to the universal scale factor shortly—and when H is evaluated at the present time, one adds the subscript 0 to it: H_0 is the value of H now; it is denoted the *Hubble constant.* Hence, the exact equality $V = H_0 \times D$, mentioned earlier in this chapter, is just Equation (4) written for the present time. The subscript 0 on other quantities will have the same meaning: it will denote the value that these quantities have now. For example, R_0 is the current value of the scale factor.

Let us reconsider the approximation to Hubble's law, relation (2). In deducing it, Hubble replaced the redshift parameter by V/c, thereby unknowingly ending up with a "law" whose form is the same as the exact result (4). Because none of the redshifts he used were even close to the limit of 0.2, his straight-line fit to the data was serendipitously correct. Had redshift values larger than 0.2 been measured, however, the approximate formula (1) would be invalid, so that if speeds had been inappropriately deduced from it, Hubble may not have been able to assume that they were proportional to distances. Of course, this is uncertain, since despite his data showing much more scatter than one might initially think is consistent with a straight line, he nevertheless drew a straight line through it. What is now clearly established, however, is that the true speed/distance relation—the exact Hubble law—is a unique consequence of an expanding Universe that manifests certain symmetries.[c]

Although my emphasis so far has been on Hubble's law, this is not intended to eclipse the fundamental character of the universal scale factor R. Far from it! R plays an even greater role in theoretical cosmology than mass does in stellar phenomena. For example, R determines how energy density, mass density, black-body temperature, and wavelength each scale with distance in an expanding universe, features explored in Chapter 8.

In addition, the universal scale factor is intimately related to the Hubble parameter H. As shown by Berry (1976), Peacock (1999), Raines and Thomas (2001), and Rich (2001), H is given by the R-dependent ratio "R's time rate of change divided by R." This statement is equivalent to the equation $H = $ (time rate of change of $R)/R$.[6] Since R has been assumed to possess the dimension of length, then the dimension of its time rate of change is length over time (which itself is the dimension of speed). Therefore, the preceding ratio is both a speed divided by a length as well as an inverse time, thus conforming to the dimensions quoted earlier for h_0. That H varies in time is now seen as a direct consequence of the time-varying nature of R and its time rate of change.

These are not the only consequences of homogeneity and isotropy. Perhaps the most significant is the reduction in complexity of the equations of general relativity. These equations—of which there can be many—govern the behavior of time and of the three spatial dimensions in the presence of matter and radiation. However, they reduce to a single pair of equations when homogeneity and isotropy are valid symmetries: many become two. Most significantly, these two govern the time variation of the single quantity R.[7] Hence, in place of the three spatial dimensions,

---

[c]Equation (4) is exact only under the assumptions of uniform expansion and homogeneity (an aspect of the Universe Hubble could not have known about). As discussed later, data from the largest distances leads to a Hubble plot displaying *curvature*, evidence that the Universe is *accelerating*.

only the time variation of R is needed to provide a complete understanding of the evolution of any homogeneous, isotropic universe governed by general relativity. This result is in complete accord with the previous conclusion that in such a universe, the change in all distances and in all directions is determined solely by R. (How unlike the growth of living things on earth, which is never uniform in all directions.)

No one should doubt that the preceding statements are among the most astonishing in science. They state that full knowledge of just one quantity is adequate to describe the evolution of a particular class of universes, a class that has features in common with our own Universe. In other words, to predict whether a homogeneous, isotropic universe will begin in an explosive event, expand, contract, oscillate in time, collapse, etc, it is only necessary to determine how R will change in time.

These comments should make it clear that one of the primary goals of theoretical cosmology has been determining the time variation of R. Because the equations it obeys contain ingredients such as the densities of matter and radiation, the cosmological constant, the Hubble constant, etc, the values of these quantities will influence the character of the solution. A different R is produced for each change in one or more of these ingredients, which are typically referred to as *parameters*. The parameters function essentially as the elements of a generic cosmic recipe, so that if you wanted to "cook" a universe, you would need to know them.

Each set of parameter values specifies a recipe for a particular R, which in turn generates a different universe, distinguished by its contents and time evolution from all other theoretical universes. There can be a vast number of them. Only one can be used to predict how ours will behave, and that is the one whose parameters are equal to those of our Universe. Definitions of the parameters, their implications, and their determination are the subject of the next chapter.

7. The Parameters of the Universe

A parameter is a quantity that can take on one or more values. A child's height or weight could be one of *its* parameters. Several of the parameters that occur in a cosmological context are time varying, in analogy to a child's height or weight. The large-scale time behavior of our Universe is characterized by the parameters that determine its scale factor. They are not only measurable but over time have been evaluated with increasing accuracy. Examples are the Hubble constant H_0 and the cosmological constant.

A change in the value of any of its parameters leads to a change in the time variation of R and thus in the theoretical universe it generates; a goal of theory is to investigate the universes that result from such changes. These changes give rise to many different scenarios, some of which I shall examine after identifying the parameters of relevance. Specification of all of them should lead to a theoretical universe whose large-scale properties will be those of our Universe, particularly its behavior at both very early and very late times.

The Parameters That Determine R

The primary parameters governing the behavior of R are listed in Table 10.[a] In addition to these six, the equations for R contain the speed of light and Newton's constant of gravity. There have been speculations that these latter two quantities might have changed over time, but no firm evidence exists that either one *has* varied. As their values are already known to great accuracy, they have been omitted from the table.

---

[a] These primary parameters refer only to the behavior of R. For CMB anisotropy analysis, cosmologists use a larger set denoted the "cosmological parameters," to which I shall return later in this chapter.

Table 10. Primary Parameters of the Universe

Symbol	Name
d_M	Density of matter
d_R	Density of radiation
H_0	Hubble constant
p	Pressure
Λ	Cosmological constant
k	Curvature parameter

The first entry in Table 10, the density of matter d_M, is the mass per volume averaged over a universe. d_M is composed of two components: the contributions from invisible and unidentified *dark matter* and from normal matter (also referred to as *baryonic matter*). At present, the former is considerably greater than the latter in our own Universe. Arguments for the existence of dark matter are summarized later. Because matter density is mass per volume, and the volume of a universe will change over time, d_M is a time-varying quantity.

Radiation density d_R, the second item in the table, refers to the *energy* of radiation, but it can be converted to an equivalent mass density by dividing the energy density by c^2, where c is the speed of light. When expressed as a mass density, the current value of d_R in the Universe is about 10^{-5} times that of the matter density. Hence, the contribution of radiation at the present time is ignored in the relevant equations. Nevertheless, while our Universe is currently matter dominated, at much earlier times radiation prevailed over matter, and the Universe was then radiation dominated. It follows that there must have been a crossover time, when the two densities were equal. It is currently thought to have occurred at roughly 40,000 years after the Big Bang and is one of the events listed in the timeline of the next chapter.

Entry three is the Hubble constant. Later in this chapter, I shall describe the recent experiments that have produced accurate values for it. Item four, the pressure exerted by the contents of a universe, has not previously been introduced. Because of the assumption that galaxies are acting like points in a dilute fluid, it will play no role in my discussion of R. On the other hand, it is

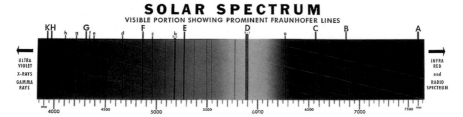

Plate 1. The solar spectrum showing Fraunhofer dark lines. The scale at the bottom is in units of 10^{-10} m, and the letters at the top of the lines identify the absorbing atoms; for example, A, B, C, D, and E, respectively, are oxygen, oxygen, hydrogen, neon, and iron. Lines A and D are each a *doublet*, that is, a pair of dark lines representing transitions to two closely spaced energy levels. (Figure courtesy of Wabash Instrument Corporation.)

Plate 2. The galaxy NGC 3949, a "Majestic Cousin of the Milky Way." The image, showing a central bulge and spiral arms, is from Hubble Space Telescope photos. (Figure courtesy of NASA and STScI.)

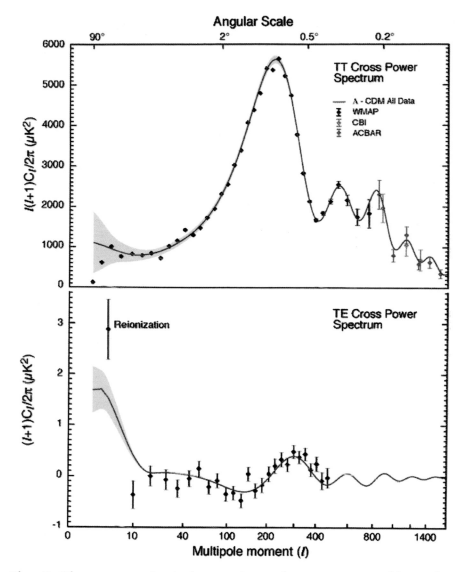

Plate 3. The upper portion is the experimental power spectrum (data points from the WMAP, CBI, and ACBAR measurements); the lower one is the analogous WMAP polarization spectrum. The solid magenta curve in the upper portion is from the WMAP analysis described in the text. The corresponding curve in the lower portion was generated using the same set of parameters. The bottom scale shows the values of ℓ; the top scale shows the angle A. (Figures courtesy of NASA and the COBE and WMAP science teams.)

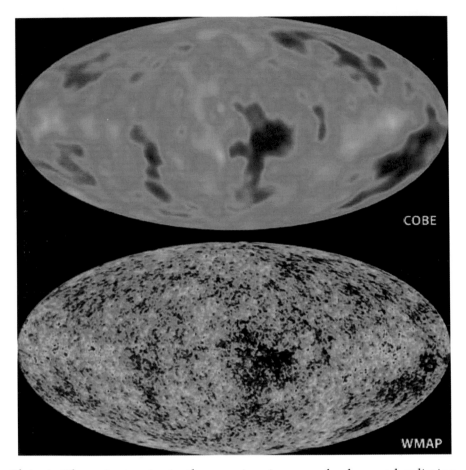

Plate 4. The anisotropies in the cosmic microwave background radiation, shown on an oval projection that represents the full sky (the plane of the Milky Way runs horizontally through the center of the oval). The false color anisotropies are the very slightly warmer (red/yellow) and cooler (blue) deviations of the CMB from the temperature of 2.275 K. The upper portion displays the results from the COBE experiment, and the WMAP results are in the lower portion. The difference in resolution between the two experiments corresponds with seeing a tree at a distance as compared with inspecting its individual leaves close up. (Figures courtesy of NASA and the COBE and WMAP science teams.)

also time varying, and because it is much greater in the early Universe, it enters the analysis of the CMB anisotropies.

The fifth entry is the cosmological constant, represented by the Greek letter Λ, the uppercase partner to λ, the wavelength symbol. Λ can be positive, negative, or zero. Introduced by Einstein to ensure that his equations described a static universe, and later denigrated by him, it is now acknowledged to be a highly significant quantity, not at all a great blunder. Its current significance is due to the recent discovery that the Universe is accelerating, a feature that has led most cosmologists and astronomers to conclude that Λ has a nonzero, positive value (*pace* Einstein). However, the value determined observationally is very much smaller—infinitesimally so—than different theoretical estimates suggest it should be (factors, possibly as large as 10^{123}(!), enter these differences; see Chapter 9).

The last entry in Table 10 is the curvature parameter k. By definition, it can take on only the three values 1, 0, or –1. Each of these values gives rise to a different geometry and thus to a different class of model universes.

Curvature

"Curved" and "straight" are adjectives that scarcely need defining when referring to everyday objects; normal experience makes the meaning of each obvious. In geometry, however, "curvature" has a technical definition related to the properties of space—which is the reason that it enters the equations of general relativity. The technical aspects are beyond the scope of this book, and my qualitative comments will deal with simple analogues and illustrative sketches.

If a three-dimensional object is not curved, it has flat surfaces. So too with *zero* curvature in geometry: $k = 0$ defines flat or Euclidean space, which, lacking curvature, is represented pictorially by a plane surface, as in Figures 18 and 23a. Let me stress here that Figure 23a is an analogue and is not meant to imply that a $k = 0$ universe is two-dimensional. If zero curvature were to charac-

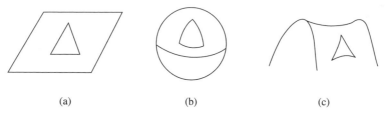

(a)   (b)   (c)

Figure 23. Analogues of the types of geometry corresponding to the three kinds of curvature. (a) Euclidean or plane geometry, illustrating zero curvature, wherein the sum of the angles in a triangle drawn on the flat surface is exactly 180°. (b) Spherical geometry, illustrating positive curvature, wherein the sum of the angles of a triangle drawn on the spherical surface is greater than 180°. (c) Hyperbolic geometry, illustrating negative curvature, wherein the sum of the angles of a triangle drawn on the saddle-shaped surface is less than 180°.

terize our Universe—and on the largest scales it does—then Euclidean geometry would describe it.

A nonpractical method of determining if the small-scale geometry of the earth's surface is Euclidean—that is, whether *locally* the curvature is zero—consists in constructing a triangle somewhere on the surface and then measuring and summing the values of its three angles. If the sum were to equal 180°, it would mean that the space is flat in the vicinity of the measurement. This procedure lacks practicality because it is so difficult to measure angles to sufficient accuracy using ordinary instruments.[b]

An important feature of expanding universes is that expansion alters the definition of distances, and this is true even when the curvature is zero. This feature will play a role in estimating the size of the visible portion of our Universe.

---

[b]However, just such an attempt to determine curvature on the earth's surface was made in the early 19th century by the German mathematician Karl Friedrich Gauss, who derived the mathematical formula for the curvature of an arbitrary surface. Gauss, involved in a land survey, was unable, with the instruments then available, to detect any effects of the earth's intrinsic curvature. With modern satellite technology and extraordinarily accurate instruments, is it possible to detect certain measurable effects of curvature on and above the earth's surface. See note 1 to Chapter 6 and the books by Berry (1976) and Webb (1999) for comments and mathematical formulae.

For $k = 1$, the space is said to have *positive curvature* everywhere; its geometry is referred to as "spherical." A convenient analogue for this case is the surface of a sphere, Figure 23b. Some of the identifying features of this surface are the following: the shortest distance between two points—a *geodesic*—is the great circle connecting them; no great circle can be drawn parallel to a given one, since the two will always intersect; and the sum of the angles of a spherical triangle is greater than 180°.

While spheres are familiar, and to some extent so is their geometry (e.g., great-circle airline routes), the last case, $k = -1$, may be unfamiliar to many readers. Here, space is *negatively* curved and its geometry, denoted *hyperbolic*, is exemplified by the surface of a saddle. For this type of surface, the sum of the three angles of a hyperbolic triangle add to less than 180°, and more than one line can be drawn parallel to a given one. Figure 23c illustrates hyperbolic geometry, although only near the seat of the saddle is the surface analogous to a $k = -1$, homogeneous, isotropic universe: farther away, the surface itself is not uniform, and therefore cannot be either homogeneous or isotropic [see Harrison (2000) for further comments].

Each of the three values of the curvature is associated with a different type of space and therefore with a different class of universe. Universes with positive curvature are referred to as *closed*, whereas those with nonpositive values are labeled *open*; additionally, spaces with $k = 0$ are conventionally denoted *flat*. Inspection of Figure 23 shows the rationale for this nomenclature, since for any given radius the spherical surface is finite in size (and therefore bounded), whereas the surfaces of the plane and the saddle can be extended indefinitely.

Qualitative Behavior of Universes with Different Values of k and Λ

The pair of equations governing the time behavior of the scale factor R is named for Aleksander Friedmann and Georges Lemaitre, and the universes generated by R are often referred to as Friedmann—Lemaitre (F-L) universes.[1] A detailed knowledge of the time variation of parameters such as d_M and d_R is required to obtain an exact, quantitative solution of the F-L equations. However, these equations can be used to draw qualitative conclusions about the universes R generates, without knowing how d_M and d_R vary with time. [If you want more details than I give below, you might try the books by Berry (1976) or Harrison (2000).]

The simplest case occurs when the cosmological constant Λ is zero and matter dominates over radiation. In this situation, all universes begin with a Big Bang; their ultimate fates will depend on the value of k. For a matter-dominated, closed universe ($k = 1$), R expands to a maximum value, say R_{max} at time t_{max}, and then decreases back to zero in a time $2t_{max}$. This closed universe ends its existence in the reverse of a Big Bang, a situation often referred to as a "Big Crunch." It is the gravitational attraction due to matter domination that stops the expansion at a maximum value of R in exactly the same way that the earth's gravity stops the motion of a ball thrown upward and then causes it to fall. Just as the ball falls to the ground, this type of universe eventually collapses into a very hot, high-density soup, just as it began.

With Λ still zero, but $k = 0$ or $k = -1$, space can be indefinite in size, as suggested by Figures 23a and 23c. In these universes, expansion will continue forever, and they end not in a bang but in a whimper, as Harrison (2000) has commented. If at some time their galaxies are emitting radiation, these two $\Lambda = 0$ classes of universe will become dark (because of the finiteness of stellar lifetimes) and extremely cold, with their absolute temperatures ultimately reaching zero. These two classes of universe differ in the final recession speeds of their cold galaxies: in the former case it is zero, whereas in the latter one it is nonzero. The explanation of this difference lies in the role played by gravity in the two cases: when $k = 0$, gravity eventually slows the expansion to zero speed at infinite separations but cannot do so for a $k = -1$ universe because of the intrinsic structure of a hyperbolic space. In each case, the associated universes will expand forever.

The scenario produced when Λ is negative is similar to that for the $\Lambda = 0$, $k = 1$ situation just described: all possible universes begin with a Big Bang and end in a Big Crunch collapse. This behavior is caused by the negative value of Λ, which acts like an *attractive* force. It reinforces the attraction exerted by gravity, which the recession due to expansion cannot overcome, leading to an eventual reversing of the expansion. Were such a universe to exist, its current age would be less than $1/H_0$, the inverse of its Hubble constant.

Because observational evidence very strongly suggests that our Universe belongs to it, the class with a positive value of the cosmological constant is the most pertinant one. Positive Λ is equivalent to a *repulsion*, or, as is sometimes said, to negative or *antigravity*. However, the time evolution of R (and the universes it generates) depends on the value of k.

For k positive and Λ *less* than a certain critical value, two different scenarios can arise. In one, the universes begin not with a Bang but with a positive value of R (and thus with a finite size), shrink to a smaller size, and finally expand indefinitely. In the other scenario, the universes begin with a Big Bang, reach a finite size, at which point the expansion reverses, and eventually they collapse in a Big Crunch. Finally, for Λ equal to or greater than the critical value, all the universes start with a Big Bang and expand forever.

The behaviors just outlined occur when both Λ and k are positive. In the other positive-Λ cases, k is either 0 or –1. All such universes start in a Big Bang and undergo an indefinite expansion whose time evolution eventually becomes exponential.[2] The exponential growth of R is related to the behavior of the parameters in the F-L equations. As time passes, the volume of a universe will expand, an increase that renders all parameters in the F-L equations negligible in comparison with Λ. Λ ultimately dominates because the other parameters, which depend inversely on the increasing volume, become smaller and smaller as the volume of the universe grows larger and larger, whereas Λ remains constant. Hence, the surviving parameter is the cosmological constant, which, acting like an antigravity repulsion, accelerates the speed of recession exponentially. Evidence for an acceleration of the expansion of the Universe was observed in the late 1990s, which is the reason for my claim that our Universe is characterized by a positive value of Λ.

These analyses describe the time behavior of the various classes of homogeneous, isotropic universes, the evolution of which is that of R. This evolutionary aspect is the basis for the claim that the expansion of a homogeneous, isotropic universe is an expansion of space itself. For, every direction in such a universe scales with R, and as R increases, so will all lengths. But, space is measured by the values of the lengths it encompasses. The claim then follows from the connection between length and R: the Big Bang is not an expansion *into* space, it is an expansion *of* space.

The Curvature Parameter and Relative Densities

Although there are only three values for the curvature parameter, it should be clear that the task of determining k directly is impossible: you cannot "see" enough of the Universe to trust any measured value, even if the curvature could be measured with sufficient accuracy. Thus, while one of the most sublime questions concerns the geometry of our Universe, indirect methods provide the only hope for answering it.

The role played by indirect methods is a recurrent theme of this book. Just as with distances, theory plus indirect measurement not only determine k, but they have also unraveled many of the mysteries of the Universe, including values of its parameters. The theory portion of the indirect procedure used to determine k is the simplest, and I describe it in this section, after introducing the contemporary notation used to express the parameters.

The notational change followed the realization that certain of the quantities contained in the F-L equations can be combined to form a new density parameter, one that has played a significant role in cosmological analyses. The new parameter, referred to as the *critical density*, is denoted by the symbol d_c. It has become the unit in which the other densities are expressed, thereby playing the same role in cosmology as the astronomical unit AU does in astronomy. I will explain soon why d_c is called the "critical density."

In contrast to d_M and d_R, d_c is independent of the volume: even though its dimension is necessarily mass divided by length cubed, the length cubed part is not the volume of the universe. The definition of the critical density contains H_0, which means that once a value for the Hubble constant has been determined, d_c can be evaluated. Here is yet another motive for accurately measuring H_0.

The value of the Hubble constant that I use in this book (derived from measurements described later in this chapter) is $H_0 \cong 71\,\mathrm{km/sec}$ per Mpc. It leads to $d_c \cong 1.37 \times 10^{11}\ M_{Sun}/(Mpc)^3 \cong 0.92 \times 10^{-26}\,\mathrm{kg/m^3}$. This tiny quantity corresponds to about five protons per cubic meter and is roughly 100 times the estimated density of luminous matter (Table 7). One reason for choosing d_c as the new unit is that it and the other densities all have exceedingly small values (in contrast, say, to the huge value of the Mpc).

When expressed in terms of the new unit, d_c becomes 1, just as the earth–sun distance becomes 1 in units of AU, Table 2. Unlike Table 2, however, cosmologists and astronomers do not simply refer to the other densities in terms of a numerical value (that is, the amount of d_c); they have introduced the new symbol Ω for these numbers. Ω, the capital Greek letter omega, is a *relative density* (it is relative to d_c because it is expressed in units of d_c). As such, Ω is a pure number like 1 or 0.4; there is no unit attached to it. Each of the Ωs is subscripted to identify the density to which it refers: Ω_M, Ω_R, and Ω_Λ are, respectively, the relative densities of matter, of radiation, and of the cosmological constant. They are the new symbols for the old parameters d_M, d_R, and Λ. Ω_M and Ω_R are time varying, but there is as yet no conclusive evidence for a time dependence of *our* Ω_Λ.

The curvature parameter for our Universe can be calculated from a relation that involves the Ωs, one that is obtained by choosing the time to be *now* in one of the two F-L equations. Since Ω_M and Ω_R are time dependent, their current values are indicated by adding the subscript 0 to the time-varying quantities (e.g., Ω_{M0}, etc.). Omitting both the negligible value of Ω_{R0} and a proportionality constant, the relation of interest is

$$k \propto [\Omega_{M0} + \Omega_\Lambda - 1]. \tag{5}$$

This proportionality shows that k is not an *independent* parameter: its value is determined by the values of Ω_{M0} and Ω_Λ. As I have indicated, the acceleration of the Universe means that both Λ and

Ω_Λ, like Ω_{M0}, are *non-negative*. Because of this, the minus sign in the square bracket of (5) is crucial to establishing whether k will be 1, 0, or –1.

Cosmologists refer to the sum $(\Omega_{M0} + \Omega_\Lambda)$ as the *total density* and represent it by the special symbol Ω_0: $(\Omega_{M0} + \Omega_\Lambda) = \Omega_0$. Using this new symbol in the proportionality (5), it becomes the slightly simpler expression

$$k \propto [\Omega_0 - 1]. \tag{6}$$

Suppose measurements lead to the result $\Omega_0 = 1$. This then implies that the square bracket vanishes, k is zero, the geometry is Euclidean, and our Universe is flat (on the largest scales). In terms of actual densities, zero curvature means that the equality $(d_{M0} + d_\Lambda) = d_c$ is satisfied. On the other hand, if $(d_{M0} + d_\Lambda)$ is greater than d_c (or, equivalently, if Ω_0 is larger than 1), the curvature parameter is positive and the geometry is spherical, whereas the geometry is hyperbolic if $(d_{M0} + d_\Lambda)$ is less than the critical density.

These comments underlie the name *critical density*. It was originally introduced to characterize the behavior of $\Lambda = 0$ universes. When their mass density is greater than the critical density, gravity will stop the expansion and then reverse it, leading to a collapse in a finite time. But when their mass density is equal to or less than the critical density, the expansion will go on forever, ending with either a zero or a nonzero speed at infinite separation, respectively. d_c is "critical" because any value of d_{M0} greater than it will lead to a closed ($\Lambda = 0$) universe. The phrase *critical density* also applies in the case of nonzero Λ: any value of $(d_{M0} + d_\Lambda)$ greater than d_c will again lead to a closed universe.

The Early Universe and the CMB

While the "simplicity" of the preceding analysis lies in the proportionality (6), k can only be determined by measuring the Ωs (and also H_0). That, of course, requires the use of indirect methods. Two of the ways in which Ω values have been deduced are from luminosities of type Ia supernovas and from the CMB anisotropy, each then to be followed by the appropriate theoretical analysis. I have already described the mechanism underlying type Ia supernovas and will now do the same for the CMB and its anisotropy. In this section, I consider the phenomena in the early Universe that led to the CMB; in the next, I shall indicate how the parameters are obtained from the anisotropy.

The "early" Universe is the time from roughly 20 minutes to approximately 379,000 years after the Big Bang. In that interval, the diameter of the visible Universe increased by a factor of 30,000, going from about 300 pc to roughly 9 Mpc, numbers based on the size estimate from the last section of this chapter and the timeline of the next. During this period of expansion, the Universe consisted of the very weakly interacting neutrinos, plus photons, electrons, positively charged light nuclei (mostly protons), and nonluminous, dark matter.

Dark matter! It doesn't shine and no one knows what it is, so could it be like the little man that wasn't there?[c] The answer is No, it really is there. So, how have astronomers inferred its existence? The first hint came from estimates of the masses of clusters of galaxies, estimates made by the Swiss-American astronomer Fritz Zwicky in 1933. He concluded from his measurements that galaxies could be gravitationally bound together in clusters only if they contained nonluminous matter. However, his measurements were approximate, and the matter remained unresolved.

The quest was resumed in the early 1970s by the American astronomer Vera Rubin, who was studying individual spiral galaxies. Spiral galaxies rotate, and their rotation speeds are calculated from the Doppler shifts of the radiation emitted by individual stars or clouds of gas. Rubin and collaborators found a puzzling result when they measured the radiation emitted by excited atoms of hydrogen and helium. Going outwards from the galactic center, the deduced rotation speeds built up to a maximum and then remained approximately constant out to the ends of the spiral

---

[c]"Yesterday upon the stair, I saw a man who wasn't there . . ." Although the composition of dark matter is unknown, there are various conjectures as to its nature; some are reviewed in the last chapter.

arms. This was puzzling because the galactic masses, previously inferred from the amount of luminous matter present, were insufficient to produce such constant speeds. Instead, the inferred masses required the rotation speeds to decrease with increasing distance from the center. The only reasonable way to account for the constant rotation speeds was to revive Zwicky's assumption: additional, nonluminous mass, what we now call dark matter, must be present in spiral galaxies.

Doppler shift measurements made some years later on stars in spiral galaxies confirmed these results and also the inference that the dark matter was distributed in a spherical halo surrounding the spiral galaxy. The amount needed to account for the magnitude of the rotation speeds is roughly *five to six times* as much as exists as normal or *baryonic* matter. That is, dark matter makes up about 85—90% of a spiral's mass while luminous, baryonic matter contributes about 15—10%, numbers that are confirmed by the CMB anisotropy analysis discussed in the next section. (Baryonic matter is the everyday stuff encountered terrestrially and in the cosmos: it can radiate. *Baryon* itself is a generic word for a proton or a neutron and also for a *quark*, their basic constituent.)

Although their masses are less straightforward to determine, the results from studies of elliptical galaxies and galactic clusters are similar to those from spirals. On the purely theoretical side, calculations have shown that the observed galactic structures can be obtained from the density inhomogeneities in the early Universe only if a significant component of the gravitating matter is dark. Taken all together, the evidence for dark matter is very strong.

Since there are two types of matter, it follows that the relative mass density Ω_{M0} is the sum of two components, one arising from baryonic matter and denoted by Ω_b, the other from dark matter and denoted Ω_{dm}: that is, $\Omega_{M0} = (\Omega_{dm} + \Omega_b)$. Despite the redundancy, each of these new densities—Ω_b, Ω_{dm}, and Ω_0—are included in the *set of redefined parameters*, displayed in Table 11. Analysis of the CMB anisotropy together with other data yields values for these parameters, plus others described in the next section.

Table 11. Redefined Parameters of the Universe

Symbol	Name
Ω_b	Relative density of baryonic matter
Ω_{dm}	Relative density of dark matter
Ω_{M0}	Relative density of matter $(\Omega_b + \Omega_{dm})$
Ω_Λ	Relative density of Λ
Ω_0	Total relative density $(\Omega_{M0} + \Omega_\Lambda)$
Ω_R	Relative density of radiation
H_0	Hubble constant
p	Pressure
k	Curvature parameter

Although dark matter is very important gravitationally, its role in the early-Universe scenario I am considering is quite different than that of baryonic matter. During the time interval in question, the charged particles, viz., the electrons, protons, and the light nuclei, formed a state of matter denoted a *plasma*. The photons interacted strongly with—they were "coupled" to— the plasma particles but were *not* coupled to the dark matter, which, being nonluminous, does not interact with radiation. Dark matter therefore enters the considerations only through its gravitational interaction with the plasma particles.

Because of their mutual interactions, the plasma baryons and electrons were largely in equilibrium with the photons at the relevant blackbody temperature. This does not mean, however, that the baryon mass density was uniform: it was not, being very slightly disturbed from uniformity because of tiny quantum fluctuations that originated at much earlier times. These perturbations were driven gravitationally by the dark matter, and the overall effect was a tendency for slight baryon mass clumping. This clumping led to a slight increase in the gravitational force acting on the baryons, which caused a tiny compression of the baryon plasma and an increase in its density.

As the compression took hold it was opposed by an increase in the non-negligible radiation pressure of the photons (d_R is not negligible in the early Universe). The increase in radiation pressure pushed the baryon clumps apart, thereby decreasing both the baryon density and the radiation pressure. Decreased radiation pressure then led to increased gravitational clumping, leading again to a compression of the plasma, followed by another increase in radiation pressure that pushed the clumps apart. A baryon compression—expansion cycle was therefore set up.

In analogy to the compression and expansion of air molecules that give rise to sound waves, the compression and expansion of the baryonic matter generated "acoustic" waves in the plasma. Their wavelengths were equal to either the diameter of the Universe at that time or to whole-number fractions of it (one half, one third, one fourth, etc.).

These acoustic waves, whose properties are determined by the inhomogeneities in the baryon mass density, fed in part by gravitational coupling to the dark matter, are predicted to be precisely mirrored in the blackbody radiation of the photons. The photon energies experienced slight variations, which resulted in slight deviations from a pure blackbody spectrum and thus in temperature variations: over-dense portions were slightly hotter, whereas under-dense ones were slightly cooler. These deviations are the anisotropies in the CMB. They changed as the Universe expanded but were frozen in their currently observable form when the temperature decreased enough to allow sufficient numbers of protons and electrons to form hydrogen atoms, a process referred to as *recombination*. Recombination occurred at approximately 379,000 years after the Big Bang (see Chapter 8).

The anisotropies were frozen in their current form because the photons lacked sufficient energy to interact with (excite) the recombined hydrogen atoms. Such *decoupled* photons formed the CMB. It has traveled almost unimpeded as nearly pure blackbody radiation with tiny anisotropies that are mainly characteristic of the acoustic wave properties of the Universe at 379,000 years.

This scenario is a broad outline of the current paradigm. It achieved paradigmatic status because experimenters were able to establish the nearly perfect blackbody nature of the CMB. The experiments began in 1989 with the launching of the cosmic background explorer (the COBE satellite experiment).[3] To demonstrate the presumed blackbody character, it was necessary to measure the CMB at wavelengths in the far-infrared, where only a few data points then existed.[4] The instrument used for this purpose was denoted the far-infrared absolute spectrometer (FIRAS).

The FIRAS team, led by the American John Mather, reported its first results in 1991: the new data points, when added to the earlier ones, fell perfectly on a blackbody curve corresponding to an absolute temperature of 2.726 K, conclusively proving that the CMB was indeed blackbody.[5] Later experiments revised this temperature to 2.725 K (!). Apart from the immediate vicinity of warm or hot bodies such as stars, planets, glowing gas, etc, 2.725 K is the temperature of the Universe at the present time.

Analyzing the CMB Anisotropy

Theorists realized early on that the parameters would be hidden in the tiny anisotropy of a nonuniform CMB, and so COBE carried out a second experiment designed to look for and measure anisotropies. Determining the current value of the parameters is not the only motive for measuring and analyzing the anisotropy. Another is the expectation of verifying predictions from a theoretical conjecture known as *inflation*, one that is beginning to achieve paradigmatic status.

Introduced in the 1980s, inflation postulates that during its extremely early history the Universe experienced a gigantic increase in size (it "inflated") in an exceedingly short time interval. Among inflation's predictions are that the Universe is flat and that the CMB anisotropy will display certain characteristics, some of which are identified later in this chapter. I will postpone examining the theoretical concepts until Chapter 9.

An anisotropy will be present in the CMB when w_λ, the total photon energy at wavelength λ, differs from the energy u_λ that 2.725 K blackbody radiation will have at that same wavelength. The energy w_λ will be either greater or smaller than u_λ, so that its deviation from u_λ will be either positive or negative.

But, whatever the sign of the deviation, its existence means that the observed photons correspond to a blackbody curve having a temperature T that differs from 2.725 K. T will be either higher or lower than 2.725 K; the difference between it and 2.725 K measures the anisotropy at wavelength λ. Since the anisotropies are roughly one part in 100,000 (see the next section), the temperature differences are about 0.00003 K. It is the ability to determine such small differences that partially characterizes the new era in cosmological measurements. These tiny temperature differences mean that measurements made from different parts of the sky can map the slightly hotter or colder spots; such maps have been constructed (see Plate 4).[6]

In the COBE experiment that first revealed the anisotropy (which I describe in the next section),[d] the CMB was measured from pairs of locations in the sky separated by fixed angles. Later anisotropy experiments proceeded in the same way, though the minimum separation angles became smaller. The range of separation angles used in an experiment is determined by the instrumentation, and in general they lie between minimum and maximum values, with the maximum in some cases being the whole sky. Individual experiments usually measure the CMB at several different wavelengths, which, like the angular scales, can vary from one experiment to another. At each of the chosen wavelengths, and over the range of angles, a positive or negative value of the temperature anisotropy is obtained.

Measuring pairs of temperature deviations makes CMB anisotropy experiments utterly different from standard ones in astronomy. In the latter case, a detector is aimed at one direction in the sky, say at a star whose luminosity is to be determined. If the detector can detect photons of only one wavelength or over a small range of them, and if the distance to the star is known, then the luminosity for the single wavelength or the small range of them is determined. For instance, the X-ray luminosity might be measured. If the full range of wavelengths can be measured, as in the case of the sun, then the total luminosity will be determined. Even if the emitter is a galaxy, and even if its radiation undergoes gravitational lensing, the detectors are essentially aimed at only a very limited angular region around one direction.

The CMB, however, is everywhere, and so there is no reason to select a particular direction from which to measure it, in con-

---

[d]A measurement in 1976 had found an anisotropy, which was soon realized to be a result of the earth's motion, an effect mentioned in Chapter 2 that must always be taken into account.

trast with the preceding situation. Instead, theoretical analysis showed that by measuring the temperature anisotropy from *pairs* of directions separated by a known angle, and then repeating the measurements by changing the separation angle, the resulting pairwise data sets would lead to the desired information. And even though the information would be hidden, further analysis could reveal it.

Thus was born the initiatives to measure the anisotropies from two different directions, the angular separations of which would be changed during the experiment. For example, initially one photon detector could be directed toward Rigel, the other toward Betelgeuse. Then the direction in which one of the two detectors pointed would be altered, say the one toward Rigel, the other direction remaining the same. The new angle between the detectors would be either larger or smaller. It is crucial for the eventual extraction of parameters that both the separation angle A and the two temperature anisotropies are known and recorded. The increments through which the separation angle ranges, as well as its minimum and maximum values, depend on the equipment used in the experiment. If the experiment can take data at more than one wavelength, the full procedure is repeated for each of them.

Before any attempts are made to extract parameters, the data must be manipulated. This occurs in several steps, the final one being the creation of a *power spectrum*, the entity from which the parameters are deduced. To obtain a power spectrum, the two members of each pair of temperature anisotropies measured at a given separation angle A are multiplied together, and then a plot is made of the values of these products *versus* the angle A. Such a plot displays graphically how the anisotropy products vary with A, just as a graph of stock market indicators shows how average stock prices have varied over a time period.

The final three steps in the process of extracting parameters are more complex: they involve two mathematical procedures that I will describe only cursorily, plus another graph. In the first of these two procedures, the preceding graph of anisotropy products *versus* angle is fitted by a curve that uses certain mathematical quantities to express the dependence on angle A. Each of these angle-dependent quantities—their technical name is a *Legendre polynomial*—is indexed by an integer usually denoted by the letter ℓ (not apparent luminosity!).[7] Finally, using technical procedures too complex to describe here, the experimental power spectrum is generated from the Legendre polynomial curve that has been fitted to the graph of products.[8]

This massaging of the data is done not only because it is the template for deducing parameter values, but also because the acoustic wave scenario (in the early Universe) predicts that the power spectrum will display a series of peaks and valleys as its values are graphed against the integer ℓ. The first peak will correspond to the acoustic-type oscillation with the longest wavelength; succeeding ones to the shorter wavelength oscillations. A sketch of a power spectrum is shown in Figure 24.

The power spectra generated from experiment *do* show such peaks, and the second of the mathematical procedures mentioned above leads to the extraction of parameter values. It is done by creating theoretical power spectra. This is a procedure simple in concept but less so in practice.

A theoretical power spectrum depends on the values of certain unknown constants in the same way that the scale factor depends on the parameters of the universe. Those that enter the generation of the theoretical power spectrum are known as the *cosmological parameters*, mentioned in the footnote on page 141; they include the parameters of Table 11. The mathematical pro-

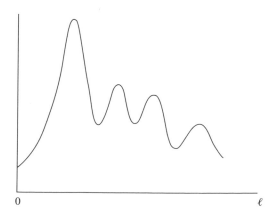

Figure 24. A sketch of a generic CMB power spectrum that has four peaks. Plotted in the vertical direction are ℓ-dependent entities that are derived from the curve fitted to the graph of products *versus* angle.

cedure that generates a power spectrum involves guessing or estimating initial values of the cosmological parameters, calculating
the theoretical power spectrum based on them, and then comparing it with the experimental one. If the comparison is not good,
the process is repeated after changing one or more of the cosmological parameters until an excellent fit is obtained. Since there
are 13 of these parameters, the game of guessing/estimating them
can be complex.

Unfortunately, an excellent fit between theory and experiment will not necessarily select the correct parameters: the procedure just outlined is, in fact, not unambiguous. Ambiguities can
arise from the choice of data sets used in creating the power spectrum, from the theoretical framework used in the analysis, and
from the choice of initial values used in the first theoretical calculation. Because of this, there is as yet no "correct" set of constants or parameters of our Universe. This is not a major drawback
as long as a qualitative understanding is desired, and that of course
is a goal of this book.

Nonetheless, the differences in extracted parameter values
can be significant. First, they produce different values of the
parameters of Table 11. At a deeper level, since the cosmological
parameters enter the theoretical description of the density fluctuations in the early Universe that eventually gave rise to the CMB,
changes in them mean changes in the theoretical model of the
early Universe. This situation is a facet of current research, and
your understanding of how the Universe is being comprehended
would be incomplete if I did not include this important feature.
Indeed, it is an aspect of cutting-edge science, especially one where
theory and experiment are still coming to grips with basic questions. (A discussion of the cosmological parameters themselves
lies outside the scope of this book, but should you want to learn
more about them, I suggest either typing that phrase into your
favorite Web browser or trying to access the "Review of Particle
Properties" issues of *Physics Letters* and *Physical Review* listed in
the Bibliography.)

In the next two sections, I will describe some of the experiments that have led to the current understanding, including a third
method for deducing parameter values. But before doing so, I
remind you that in addition to providing values of the cosmolog-

ical parameters, analysis of the power spectrum and other data will confirm or disprove some of the predictions of inflationary theory; for example, inflation's prediction that the Universe is flat will be verified if the position of the first peak in the power spectrum is found to lie at a specific position (identified later).

Many significant CMB measurements have been carried out since the early 1990s; those of another type since the late 1990s. In the following section I will review, in chronological order, some of the results obtained prior to those of the Wilkinson Microwave Anisotropy Probe (WMAP) experiment, including paradigm-changing ones from type Ia supernovas.

Pre-WMAP Measurements

The COBE experimental team that searched for anisotropies in the CMB, led by the American George Smoot, used an instrument known as the differential microwave radiometers (DMR). It consisted of three pairs of microwave detectors, whose antenna "horns" could observe regions in the sky varying from 7° to the full sky. Each pair took data at different infrared wavelengths. In all such measurements, the range of angular scales depends on both the optical design of the experiment and how much of the sky is observed. One limitation of the DMR horns was their inability to see radiation sources closer together in angle than 7°.

In 1992, DMR yielded the first evidence that the background radiation contained the anticipated anisotropies. They ranged from roughly 1 part in 100,000 to several parts per million and were found over the complete angular range available to the instruments.

The 1991 and 1992 COBE results were like a one-two punch: a community that had anticipated them began preparing for and initiating a series of increasingly more precise measurements. Of special interest was CMB data from angular separations smaller than 7°, for theory had shown that only such data could lead to parameter values.

The new anisotropy experiments that followed soon after publication of the DMR results included both earth-based ones and a few that used balloons to carry the instruments aloft. The former were successful; initially, the latter were not. Although the results of the early earth-based experiments contained large experimental uncertainties, the combination of their data with those from DMR yielded a small portion of the full power spectrum, enough to suggest the existence of a first peak. Its existence was confirmed by the earth-based experiments of a group at Princeton University led by the American astronomer Lyman Page; its position strongly suggested a flat (an $\Omega_0 = 1$) Universe.

As you might have guessed, this finding stressed the need to perform measurements with the smaller separation angles and the greater precision needed to see other peaks in the power spectrum. To this end, new balloon experiments were launched. However, before they could provide additional power-spectrum information, the two type Ia supernova collaborations (see page 113) announced a result of paradigm-shifting significance. Analysis of their data led to the conclusion—in early 1998—that the Universe was not simply expanding, it appeared to be accelerating!

From the behavior of the different classes of universe that I outlined earlier in this chapter, the presence of an acceleration implies that our Universe is governed by a positive cosmological constant, an implication the two collaborations did not fail to recognize. Their result brought Λ into the limelight from its backstage role as a theorist's invention (or blunder), allowed but not previously suggested by experiment.

The conclusion that the Universe is accelerating was also remarkable because the evidence available to the collaborations just a year earlier had suggested that the expansion was decelerating. This result was consistent with the belief then held that we lived in a mass-dominated Universe for which $\Lambda = 0$. Another year's effort provided the data that led to an improved Hubble plot of type Ia supernova brightness *versus* redshift z. At the larger values of z, the data lay on an upward curving line. This upward curve was the evidence for an acceleration of the expansion and the presence in our Universe of a positive cosmological constant.

In addition to reaching this latter conclusion, the supernova collaborations used their results to estimate values for Ω_Λ and Ω_{M0}. First, they displayed their data on a plot of Ω_{M0} *versus* Ω_Λ. Then, by invoking the inflation scenario and its prediction that the Universe is flat ($k = 0$), they found a good fit to their data for the values $\Omega_{M0} \cong 0.25$ and $\Omega_\Lambda \cong 0.75$. (If $k = 0$, then relation (6) requires that $(\Omega_{M0} + \Omega_\Lambda) = 1$, which is approximately satisfied by the preceding numbers.) These estimates were the first obtained experimentally and were the backdrop against which the results of the balloon experiments would be compared. The value of Ω_{M0} was especially intriguing, since it was consistent with the widely held belief that most of the mass in the Universe is composed of dark matter.

By 2000, the analyses of results from the first successful balloon experiments were presented. The data came from two acronymic collaborations. The first to report was denoted BOOMERanG, then led by the Italian Paulo de Bernardis and the American Andrew Lange. The announcement of their findings was quickly followed by that of the MAXIMA collaboration, led by the American Paul Richards. Both collaborations were able to measure separation angles as small as 10 minutes. That is, their *angular resolution* was one sixth of a degree, a 10-fold increase in resolution over DMR. This enabled them to create a power spectrum from their combined data that could contain at least two peaks. Two were reported, as was the possibility of a third one. The position of the first peak, at the value of the power spectrum parameter $\ell \cong 250$,[9] was consistent with that of the Princeton group, each supporting the earlier conclusion that our Universe is flat. This in turn was additional evidence favoring the inflation scenario.

For the Universe to be flat, the *total relative density* Ω_0 ($= \Omega_{M0} + \Omega_\Lambda$) must equal unity to within experimental error. The values found for it by the two collaborations were $\Omega_0 = 1.06 \pm 0.06$ (BOOMERanG) and $\Omega_0 = 0.90 \pm 0.07$ (MAXIMA). While the first peak in the power spectrum leads to a value for the total relative density Ω_0, the height and position of the second one is a measure of Ω_b (and higher ones are a measure of other parameters). The two collaborations initially found values for Ω_b that were larger than had been obtained from primordial nucleosynthesis calculations; that is, from the amount of each of the light nuclei formed in the early Universe, especially the amount of deuterium. This implied that either the nuclear physics that had been used to perform the primordial nucleosynthesis calculations was fundamentally in error or that the inflation scenario, partially supported by the power spectra, and highly attractive to its adherents, was a flawed conjecture. Neither possibility was a desirable one.

These undesirable possibilities were rendered moot by two sets of results announced in 2001. The first was the reporting of a new measurement and analysis of the CMB anisotropies by the Degree Angular Scale Interferometer (DASI) experimental collaboration, led by the American John Carlstrom. It found clear evidence of a third peak in the experimental power spectrum, and a height of the second peak consistent with expectations (see note 9). Their extracted value for Ω_b was 0.042 ± 0.008, in excellent agreement with that from primordial nucleosynthesis, while their value for Ω_0 was 1.04 ± 0.06. The latter number provided continuing evidence supporting the prediction of a flat Universe.

The numbers quoted for the Ωs depend on the value of the Hubble constant H_0. The one used by the DASI team was 72 ± 8 km/sec per Mpc, the value determined by the Hubble Space Telescope's Key Project. It was a milestone result, as you will soon see.

In the journal article that the DASI team later published on their experiment, they included both revised and additional values for some of the parameters, as follows: $\Omega_{dm} = 0.27 \pm 0.08$, $\Omega_{M0} = 0.40 \pm 0.15$, $\Omega_\Lambda = 0.60 \pm 0.15$, and $\Omega_0 = 1.00 \pm 0.04$. To within the limits of the uncertainties, the values for Ω_{M0} and Ω_Λ were in agreement with those from the supernova collaborations and revised ones from BOOMERanG; the relation $\Omega_{dm} \cong 6 \times \Omega_b$ was maintained; and the inflation prediction of $k = 0$ was slightly strengthened. It should be evident that the advantages of measuring and analyzing the CMB anisotropies were being realized.

The year 2001 also saw a new result of quite a different kind: the Hubble Space Telescope's Key Project announced an improved value for the Hubble constant. Led by the American astronomer Wendy Freedman, the Project's initial goal was to evaluate H_0 to an accuracy of 90%, partly through the recalibration of distances to Cepheids in a number of different galaxies (the Project's Web site is listed in the Bibliography). The value for H_0 presented in their final report was an average of the values obtained from a total of five different standard candles. The averaging yielded the result $H_0 = (72 \pm 8)$ km/sec per Mpc, noted above.

This value for H_0 is to be contrasted with the much lower one of 55 and the much higher one of 100 that had been advocated in the old controversy [see the book by Webb (2001) for comments]. To gain some "feel" for these numbers, let us ignore the acceleration of the Universe and, as in Chapter 5, use $1/H_0$ as an age estimate. The ages yielded by the high, intermediate, and low values of H_0 just stated are 9.5, 13.2, and 17.3 billion years, respectively. The $H_0 = 72$ km/sec per Mpc intermediate age is consistent with both the age range of 13—14 billion years obtained from the white dwarf analysis of certain 2002 Hubble Space Telescope results[10] and the age of 13.7 billion years determined from the analysis of data from WMAP and other sources.

There are three more pre-WMAP measurements that I will describe in this part of the CMB story, but in case you're feeling a bit overwhelmed by the amount of material I've presented already, I'm going to pause for a moment to review the big picture. In other words, what is the forest whose trees you've been attempting to hug? It mainly consists of the definitions and meanings of the parameters plus the kinds of universes they produce; of the clumpiness/acoustic wave character of the early Universe; of the procedures for analyzing the CMB anisotropies; and of some pre-WMAP experiments that yielded values for a few of the parameters and have supported the inflation-based prediction that the large-scale geometry of the Universe is Euclidean ($k = 0$). The remainder of the chapter will round out this picture with the wonderful 2003 measurements and the parameter values deduced from them, along with values for the age of the Universe and the estimated size of its visible portion.

Let us now zero in on the remaining pre-WMAP measurements, whose results were reported in 2002. The experiments were those of the Cosmic Background Imager (CBI) collaboration, led by the Americans Anthony Readhead and Stephen Padin; the Arcminute Cosmology Bolometer Array Receiver (ACBAR)[e] group, led by the Americans William Holzapfel and John Ruhl; and the DASI team.

The CBI detector, designed to extend the range over which the power spectrum was predicted to display peaks and valleys,

---

[e] A bolometer is a temperature measuring device.

performed as well as the BOOMERanG, MAXIMA, and DASI instruments had (the CBI Web site is listed in the Bibliography). The CBI data tended to confirm yet another prediction of inflation, viz., that the height of each successive peak in the power spectrum should be less than that of the previous one.[11] When the CBI data was combined with that of the other three collaborations, the results indicated a total of five peaks. Use of essentially the same parameters and thus the same theoretical curve that had been employed by the earlier analyses gave an excellent fit to the combined power spectrum.

The ACBAR collaboration (Web site listed in the Bibliography) also found strong evidence for five peaks in the power spectrum. Their magnitudes successively decreased, apart possibly for peak number two relative to number three, where the error bars suggested equality of the two, just as with the CBI result. Although the parameters obtained from their analysis differed to some extent from what had been a previous consensus, their data, along with that from the CBI collaboration, has been used in the analyses described in the next section.

The DASI collaboration reported the first measurement of the *polarization* of the CMB. Polarization is an intrinsic property of light and radiation, and considered simply as a word, it will be familiar to readers who have purchased polarized sunglasses. By using polarized lenses in sunglasses, some or most of the sunlight whose polarization direction is perpendicular to that in the lens is diminished, if not eliminated, thereby reducing glare.

In 1968, the British astronomer Martin Rees predicted that the CMB should be polarized, a characteristic that was later shown to be a consequence of inflation theory. By carrying out a difficult experiment, the DASI team not only showed that the CMB *was* polarized but also that the measured polarization could be successfully fitted with a theoretical curve whose parameters were just those previously extracted from the CMB power spectra. It was an experimental and a theoretical triumph. And, as discussed in Chapter 9, while there are problems concerning the theoretical foundations of inflation that have yet to be overcome, the supporting evidence for this scenario is very strong.

WMAP, SDSS, and Beyond

The CMB experiments recounted so far have been performed on or near to the earth's surface: COBE was a near-earth satellite mission, BOOMERanG's equipment was borne aloft in balloon flights sent up from the South Pole, MAXIMA's was in a balloon launched from southeast Texas, DASI is an instrument at the coast of Antarctica, CBI is located on a high Andean desert in Chile, and ACBAR's detector is mounted on the Viper telescope, also in Antarctica.

If these sites may collectively be thought of as ones local to the earth, then by contrast, the Wilkinson Microwave Anisotropy Probe (WMAP) experiment must be designated as "far from the madding crowd": its instrumentation is located in a satellite more than 1.5 million km from earth, in a direction opposite to the sun. The satellite moves in a slightly unstable orbit that follows the earth as it revolves around the sun; the orbit needs minor adjustments every 3 months or so. Their Web site is listed in the Bibliography, and a history of the project has been published by Lemonick (2005).

The WMAP, which was launched in 1999 and whose principal investigator is the American Charles Bennett, has excellent views of the whole sky. With an angular resolution 40 times that of COBE, it has supplied highly precise data from which much information has been obtained. The WMAP collaboration announced its first results early in 2003. They had obtained them by analyzing a large data set that came from their own measurements combined with those from other astronomical/cosmological experiments, including CBI and ACBAR. The greatly improved precision led to parameter values with smaller uncertainties than had been previously obtained.

A combined WMAP power spectrum and the best-fit theoretical spectrum from which cosmological parameters are extracted is shown in the upper portion of Plate 3. The agreement is excellent. In the lower part of the figure is an experimental polarization spectrum and the theoretical curve generated from the *same* set of parameters that were used to generate the upper curve. As there were no adjustable constants used in the lower figure, the agreement between theory and experiment is seen to be very good.

Table   12. Representative   Values   of   the
Parameters*

Ω_b	0.044 ± 0.004
Ω_{dm}	0.23 ± 0.04
Ω_{M0}	0.27 ± 0.04
Ω_Λ	0.73 ± 0.04
Ω_0	1.02 ± 0.02
Ω_R	$\cong 0$
H_0	$71 + (4 - 3)$
$p^\dagger$	$\cong 0$

The units of H_0 are km/sec per Mpc.
*Combined WMAP values (see note 12).
†Pressure refers only to that from matter and radiation; the
(negative) pressure due to Λ is not being considered.

Values for the parameters of Table 11 that were extracted in generating the above curves are listed in Table 12.[12] (It is labeled "representative values" for reasons spelled out below.) Since the position of the first peak in the power spectrum strongly suggests a flat Universe ($k = 0$, $\Omega_0 \cong 1$), this conclusion has been incorporated into the WMAP analysis as well as into many other ones. From Table 12 it follows that there is a little more than five times as much dark matter as ordinary matter, a negligible (relative) density of radiation, and a density of the cosmological constant that overwhelms that of matter by roughly a factor of three.

Although the preceding numbers produced no major surprises—I shall return to them shortly—the enormous increase in accuracy led to the creation of a spectacular map of the Universe showing the hot and cold spots (the temperature anisotropies) over the whole sky. The WMAP anisotropy map is shown in Plate 4; its finer scale results are compared with the much less detailed map deduced from the COBE results. It is a stunning display of the value of improved accuracy.

These representative values do not represent the end of the quest, for many other sets have been obtained, some by the WMAP team, and none can yet be considered final. It is this general situation that led me to use the modifier "representative" in Table 12. A specific reason is that improved data will undoubtedly become available in the future. Another is that different parameter values can be and have been used to fit the data. It is therefore time to consider the ambiguity problem mentioned earlier, which I shall do using two non-WMAP reanalyses.

In the first, by O. Lahav and A. Liddle,[13] reanalysis of the combined WMAP data led to two other sets of cosmological parameters and thus two more sets of the parameters of the Universe. Like the original WMAP team analysis, only 6 out of the 13 cosmological parameters were allowed to vary; the others were fixed in advance. In both cases, a good fit was obtained to the power spectrum. Those six parameters, being the minimum needed, were later referred to as providing a "vanilla model" of the density fluctuations in the early Universe. As an illustration of the changes resulting from this reanalysis, the values obtained for H_0 are 71 ± 4 and 72 ± 5 (km/sec per Mpc). Changes in the other parameters are correspondingly small (see note 12).

The other reanalysis, which was more comprehensive because it included certain non-CMB data, was carried out in 2003 by the Sloan Digital Sky Survey (SDSS) team. Directed since mid-2003 by the American astronomer Richard Kron, SDSS began in the 1980s with the goal of determining the position and luminosities of more than 100 million celestial objects, as well as measuring the distances to more than a million galaxies and quasars. (The SDSS Web site is listed in the Bibliography.)

The multifaceted approach of the SDSS team reanalyzed the combined WMAP data with and without the addition of a new experimental power spectrum deduced from SDSS data on more than 200,000 galaxies. One of their major conclusions was that the same vanilla model (a phrase they introduced) that works so well in fitting the combined WMAP data also fits the total data set, that is, the one obtained by including their own power spectrum results (which also reduced the experimental uncertainties).

Another conclusion was that by removing the prior choices for the other cosmological parameters, a large range of parameter values is allowed. This happens whether or not the SDSS power spectrum data is added in; the ambiguous nature of the enterprise is thus reinforced. Again taking the value of H_0 as an indicator, they found it could lie between 48 and 72.5 km/sec per Mpc (I have omitted the error uncertainties).

Ambiguities are inherent in the procedure, and so I stress again that the reported variations in parameter values should not be regarded as a major drawback. To help understand why this is so, recall that parameters are implicit not explicit in the analyses. A somewhat farfetched analogy would be trying to determine the value of a grown woman's age in years and height in centimeters if you were told that the sum of the two numbers is, say, 220. It would seem reasonable to suppose that her height would lie in the range 130 to 180 cm, so that the limits on her age would be 40 to 90 years. Insufficient information has clearly been given, but the ranges might be narrowed by stating that her natural hair color contains no gray and that she is still of childbearing age. Even with this additional information, there is a range of values, just as in the parameter analyses. In the case of the parameters, there is high expectation that the values of at least certain of the parameters will lie in narrow ranges.

Because the SDSS team's preferred values—$H_0 = 70 + (4 - 3)$ km/sec per Mpc and $\Omega_{M0} = 0.30 \pm 0.04$—are not too dissimilar from those of Table 12, I have retained the latter as the representative ones. They will probably change by small amounts in time, a feature you could keep track of via occasional Web site watching. To summarize, the overall finding from the WMAP and SDSS investigations is that, to within a small percentage variation, the relative densities of the cosmological constant and of matter are roughly 70% and 30%. The former remains an electrifying conclusion, one that was not taken seriously until 1998, when the supernova collaborations announced that the expansion of the Universe is accelerating.

SDSS also reported other results, among them a vastly improved upper limit on the neutrino mass: less than (6.4×10^{-4}) × M_p, where M_p is the mass of the proton. Several members of the SDSS collaboration have also produced a new map of the Universe that maintains shapes locally but nevertheless shows distances ranging from the solar system out to 2 billion light years. Readers interested in this map can find it referred to on the SDSS Web site. And, although galactic structure and clustering are not topics dealt with in this book, SDSS's discovery of the largest observed structure in the Universe, a wall of galaxies 1.37 billion ly long, is an event worth noting.

Before turning to my final topics, the age of the Universe and the size of its visible portion, let me emphasize—even celebrate—

what I hope you have realized is a stunning conclusion: the parameter ambiguities notwithstanding, the answer to the question raised in the Introduction, of which theoretical universe most closely corresponds to ours, seems close at hand. It is the general relativity–based, homogeneous, isotropic universe whose parameters will be the final ones determined by terrestrial and satellite observations of the type I have described above. And, it will be the ultimate framework supporting detailed studies of galactic structure and formation, etc.

Bear in mind as well, that it is solely in the foregoing context that the data implies a flat Universe. But, because homogeneity and isotropy characterize the Universe only on the largest scales, it follows that no claims are being made that $k = 0$ holds locally. This should be obvious: the space around a black hole, for example, is strongly distorted, as is the space in the vicinity of galaxies, evidence for which is gravitational lensing.[14]

The Age of the Universe

Not surprisingly, the age of a homogeneous, isotropic universe governed by the equations of general relativity is given by a mathematical expression too complicated to evaluate other than numerically. (In mathematical terms, the expression cannot be evaluated analytically, though it can be simplified in certain cases.) The formula for the age is an expression that depends on H_0; on the parameters Ω_{M0}, Ω_Λ and Ω_R; and on one of the cosmological parameters. Because the period when radiation dominated over matter was relatively short, Ω_R is typically eliminated from the age formula, an estimate for which then turns out to be just the Hubble lifetime, $1/H_0$. Using the combined WMAP value for H_0, the Hubble lifetime is approximately 13.4 billion years, as stated previously, while the SDSS value of H_0 yields a Hubble lifetime of approximately 13.6 billion years.

Each estimate is consistent with the recent age range for the Universe deduced from the Hubble Space Telescope measurements on white dwarfs noted earlier in this chapter and with the exact ages of the Universe listed by each collaboration. These latter ages are 13.7 ± 0.2 billion years (combined WMAP) and 13.5 ± 0.2 billion years (SDSS vanilla model), numbers within each other's uncertainty ranges. I am using the combined WMAP value as a representative lifetime.[f]

Although the uncertainties in the latter pair of numbers are small, you should note that the numbers are derived from general relativity, which, as a "classical physics" type of theory, does not incorporate quantum effects. It is widely believed that once a paradigmatic quantum theory of gravity has been formulated, it will include general relativity as a limiting case, but such a theory might conceivably alter the current picture of the Universe, including its age and any size estimates.

In view of this mild caveat, it is worth pausing to reemphasize the premise that underlies the determination of all the quantities characterizing the Universe. The premise is that there exists a unique correlation between the general-relativity description of a homogeneous, isotropic universe and our Universe, as long as the measured cosmic parameters of the Universe are used to evaluate the relevant mathematical expressions. In particular, the statements concerning the age, history, acceleration, Big Bang origin, composition at very early times, etc, of our Universe are based on calculations using the general relativity equations of its theoretical counterpart.

Lest this last remark appears to be too abstract and theoretical, I should also stress that consistency tests as well as predictions arising from the general relativity framework have received experiment confirmation.[15] Among the latter are the expansion, the existence and properties of an isotropic, polarized, low-temperature CMB, and the numerical values of the relative abundance of certain light elements, a topic considered in detail in the next chapter. That is, the Big Bang, general-relativity cosmology used to describe our Universe has so far passed the tests put to it. One caution, of course, is that future measurements may require changes: what is reported in this book comes from the currently accepted picture.

---

[f]SDSS lists a variety of ages that result from their different analyses, ranging from 13.32 to 16.5 billion years (errors deliberately omitted).

The Size of the Visible Universe

So, how big is the visible Universe? Reliable distance measurements are obviously needed to answer this question, but as I have stressed, distances are determined from other measurements. It is clearly essential to know what is actually measured. Certainly not distances at the cosmological level! Instead, observations provide apparent luminosities, light curves, spectra/redshifts, and CMB anisotropies. Parameters, etc, are then deduced from these measurements. To obtain a size, formulas from general relativity that contain the parameters are employed, yet each of these formulas yields a different expression for distance (this is the reason that redshifts and ages, rather than distances, are usually quoted when describing type Ia supernova results).

The lack of a unique distance formula can be understood by means of the "receding finish line" analogy. Imagine a 500-meter race in which the finish line begins to move away from the runners after the starting gun is fired, and further, that information is communicated at a speed comparable to that of both the average runner and the receding finish line. Suppose that at some specified time during the race, a runner wishes to know how far away the finish line is. Any information that provides that distance would be exact only if it were communicated to the runner instantaneously (i.e., only if the communication speed were infinite), since in this situation, neither the runner's nor the finish line's position would have changed. But the finite speed of the communication signal means that both of these latter positions *would* have changed by the time the signal reached the runner, and therefore the information it conveyed would no longer be valid. Although that distance would be inexact, it is generated by one possible definition of distance. You may wish to consider other possible definitions.

This example helps illustrate the lack of a unique distance definition in an expanding environment, be it a race or an accelerating Universe. The fact that in our Universe the information is conveyed at the enormous speed of light does not negate the argument. A further complicating factor is an effect of the expansion on radiation. As remarked on by Rich (2001), in an expanding universe there should be a spread in the arrival times of successive light pulses emitted by the same source, a prediction that has been verified to within experimental error by measurements on high-z supernova explosions.

This lack of uniqueness means that the distance definition being used should be specified when interpreting data. There are five different definitions and corresponding formulas arising from general relativity [see, for example, Webb (1999)]. The two most often used are the *luminosity distance*, related to the apparent luminosity, and the *angular distance*, related to the angle subtended by an extended source. Another is the distance associated with the *lookback time*, which is the difference in time between emission of the radiation and its observation; it is the distance usually referred to when a source is said to be a particular number of light years away. Of course, by the time the radiation has been received on earth, the source is no longer at the emission point.

The mathematical expressions for all the distances involve H_0 and the Ωs, as well as the redshift parameter z. Simple formulas occur only in certain limiting situations, for instance, small z or a universe in which matter dominates and the cosmological constant is zero. Unfortunately, in trying to estimate the size of our Universe, neither of these limiting situations apply: z is not small, and Λ is not zero. Furthermore, it is not evident that the largest observable value of z has yet been measured, nor is it likely that if this does occur, it will happen soon: as I discuss in the next chapter, the larger the value of the redshift, the further back in time one is "seeing."

So, a strict answer to the question of size is that the size of the visible Universe is not precisely known. And this should not be surprising: the current value R_0 of the scale factor is unknown [see the next chapter for (non-) ramifications]. As noted above, cosmologists not only tend to refrain from estimating either the size of the observable Universe or distances to remote objects, they also refer instead to an object's age, obtained from its redshift. The time/redshift relation is explored in the next chapter.

This is not an entirely gloomy situation, however, since there is a simple argument that leads to an estimate for the size of the visible Universe, one that may not be off by too large an amount.[16] It leads to the value, inferred on pages 6 and 120, of 8500 Mpc for the size (actually the diameter) of the visible Universe. There are two steps to the argument. The first is that in an expanding universe with a positive cosmological constant there will be a distance beyond which you cannot see (i.e., from beyond which no photons will ever reach you). The analysis that underlies the existence of the limiting distance is that a nonzero Λ, acting like antigravity, accelerates the expansion speed. As a result, radiation emitters such as galaxies recede farther than they would at constant expansion speed. Eventually they are carried so far away that their light can never reach us [in technical terms they have passed beyond the relevant event horizon (see note 16)].

The second step in the argument recognizes that an estimate of this limiting distance is simply the maximum distance light can travel during the lifetime of the Universe. The maximum distance is the product of the speed of light multiplied by the lifetime, for which I will use the age estimate of $(1/H_0)$. Multiplying these two quantities yields a distance that can be used as the radius of a sphere, which, when it is centered on the earth, is denoted the *Hubble sphere*. Every point on its surface is at the limiting distance. Because of this, the surface of the Hubble sphere approximates the outer portion of the visible Universe as seen from the earth. The diameter of the Hubble sphere is therefore an estimate of the size of the visible Universe. Since the diameter of a sphere is twice its radius, the size estimate becomes twice the product of the speed of light times the age estimate $1/H_0$. Using the combined WMAP value for H_0,[g] the latter product yields the previously quoted value of 8500 Mpc.[17]

This estimate for the diameter of the visible Universe is approximately equal to 2.6×10^{23} km. It is about 6.5×10^9 times the distance to the nearest star, which is a number roughly 280,000 times bigger than the diameter of the Galaxy. To put this in perspective, were the Galaxy the size of an aspirin tablet

---

[g]If the SDSS value of H_0 is used, the size estimate reduces to roughly 8400 Mpc, a number sufficiently close to the above value not to require change. Note that since an estimate ignoring acceleration is involved in calculating the size, it would be inconsistent to use the exact ages quoted above from the combined WMAP and the SDSS collaborations: these numbers were obtained taking the acceleration into account.

(0.63 cm), the diameter of the visible Universe would be about 1.8 km.

Notice, by the way, that since the preceding argument refers only to the visible portion of the Universe, it is neutral on the question of whether the Universe extends beyond its visible portion, a point that I will consider in the last chapter. Another question concerns the expansion: will the Universe continue expanding in the indefinite future or will it stop, possibly contracting and eventually collapsing in a "Big Crunch"? Here, current theory is unambiguous: the fact that Λ is positive means that the Universe will expand forever.

8. The Early Universe

"Once upon a time..." Children's stories and fairy tales often start with these words, but they could equally well usher in a different kind of story, the history of the Universe: "Once upon a time, there was a very tiny Universe that was very, very hot, and in it lived dark matter, neutrons, protons, electrons, antineutrinos, and photons." This scenario forms part of the cosmologic gospel, whose list of early inhabitants is based on the relic background radiation, the laws of nature, and the results of astronomical and laboratory experiments. It developed from the work of Alpher, Follin, and Herman on primordial nucleosynthesis and has enabled cosmologists not only to deduce which events occurred as the Universe evolved but also, with increasing accuracy, to pinpoint the times of their occurrences.

Among the best understood of these events are ones that took place during the *early Universe*, which is the period from roughly 10^{-6} seconds to about 379,000 years after the Big Bang. A number of later events are also well understood, and I shall refer to several, but my main emphasis will be on the preceding era. I emphasize it in part because the prediction of the relative abundance of the nuclei produced during primordial nucleosynthesis, outlined below, has been verified experimentally.[1] This crucial result is one of the several pieces of evidence confirming the Big Bang scenario.

In deducing the early history of the Universe, cosmologists have an advantage over ordinary biographers: the scientific conclusions concerning intimate details of the early Universe never involve hearsay. Instead, the relevant information and conclusions are universally accepted as reliable. And, where terrestrial laws do not hold or are not fully developed, the scientific conclusions are replaced by conjecture, the topic of the next chapter. In this one, however, the journey is over firm ground.

Quantifying the contents, interactions, and evolution of the early Universe involves complex mathematical analyses, for whose qualitative description I will use the cosmic microwave background and the fact that it is almost perfectly blackbody. Its blackbody nature enters in two ways. The first can be expressed as "once blackbody, always blackbody." That is, in a homogeneous, isotropic universe, expansion or contraction cannot change blackbody radiation into any other form: it will remain blackbody. What an evolving universe *will* do is change the characteristic temperature of such radiation, which decreases when expansion occurs and increases under contraction.

The maintaining of its blackbody character leads to the second property of the CMB, that it is an extraordinary diagnostic tool. Of special importance in this context is the fact that the CMB temperature T_{ph} is inversely proportional to the scale factor R: $T_{ph} \propto 1/R$. Because of this, calculation of the time variation of R means that the time variation of T_{ph} has also been determined.[2]

This last statement implies that throughout the evolution of the early Universe, the details of the CMB—the distribution of photon energies over wavelength—can be established. It is knowledge of the photon energies that allows cosmologists to pinpoint the times when different events occurred. The procedure is simplest to describe by working backward in time through acts of mental time travel. It is the mental equivalent of running a movie film in reverse.

Traveling Back in Time *via* the CMB

In the imaginary journey of going back in time, the Universe, instead of expanding, will contract. Correspondingly, the scale factor will decrease, which causes the blackbody temperature T_{ph} to increase, a conclusion that follows from the relation $T_{ph} \propto 1/R$. As T_{ph} increases, so does the total number of photons, while the maximum in the blackbody distribution curve rises dramatically and also shifts to smaller wavelengths. (For details, see the discussion in Chapter 4 and especially Figure 12a.) As a consequence, at earlier times in the Universe there will be more blackbody photons with higher energies and fewer with lesser ones than at later times.

As I will show soon, there was a period after recombination and prior to the formation of stars and galaxies when the Universe consisted mainly of hydrogen atoms, the CMB photons, anti-neutrinos, and dark matter. Not only were these the only objects present in substantial amounts, but also there were few if any photons that could ionize the hydrogen atoms. In another words, T_{ph} was still too small for the CMB to contain very many photons with enough energy to break hydrogen up into a free electron–proton pair. (The antineutrinos were also in the form of blackbody radiation, but their blackbody temperature was much too low for them to be able to ionize hydrogen. The antineutrino background also maintains its blackbody shape as the Universe expands and is in principle detectable now, but because neutrinos interact so weakly with matter, it is not possible at the present time to do so—and it may never be.)

In the preceding scenario, hydrogen atoms and CMB photons form a noninteracting gas, since neutral hydrogen, unlike charged particles, does not absorb or scatter low-energy photons. However, by continuing the mental journey back to times prior to recombination, T_{ph} becomes large enough for the CMB photons to have sufficient energy to start ionizing hydrogen. And the further back one goes on this mental journey of time travel, the more energetic will the CMB photons become, with the result that there will be fewer and fewer non-ionized hydrogen atoms. Eventually all the hydrogen will be ionized, and the Universe will consist mainly of dark matter, protons, electrons, photons and antineutrinos.

Let me pause in this journey and reiterate: if the CMB photons were not blackbody, whose properties are completely known for all sizes of the early Universe, it would very likely be impossible to specify the times when specific, energy-based events had occurred. For example, a random distribution of photon energies would not be a viable diagnostic tool.

The preceding example is illustrative of the behavior in general: as one goes further back in time, R decreases, T_{ph} increases, greater numbers of photons have increasingly higher energies, and these then become increasingly capable of breaking up the more tightly bound microscopic systems into their constituent parts. In the end, that is, as far back as one can go prior to being speculative, the Universe becomes ultimately simple, in that no composite systems exist: there is too much energy to permit permanent formation of them.

This phenomenon is not limited to atoms. Another instance that will be encountered in the mental journey back in time involves the very light nuclei formed in primordial nucleosynthesis, for example deuterons or ^{3}He (bound systems consisting, respectively, of a neutron and a proton, and of a neutron and two protons). For photons to be able to disintegrate nuclei, the Universe must be much hotter than it was during ionization of hydrogen atoms. The reason is that the energies involved in atomic processes such as ionization of atoms or chemical reactions are roughly one millionth of those needed to disintegrate light nuclei: recall the discussion in Chapter 4 on stellar energetics. The temperatures involved are $T_{ph} \cong 3000\,\text{K}$ when recombination occurs and $T_{ph} \cong 7 \times 10^8\,\text{K}$ at the onset of light-nuclei disintegration. The combined WMAP analysis indicates that the latter event occurred when the Universe was about 200,000 times smaller than at recombination, a number that is approximately equal to the ratio of the first to the second of the two photon temperatures. (Because the combined WMAP parameters are my representative ones, then consistency will require that the combined WMAP times are the ones defining the occurrence of events in the evolution of the Universe.)

As noted at the outset of this discussion, energy plays a critical role: for ionization or disintegration to occur, the CMB temperature needs to be high enough that sufficient numbers of photons with the requisite energy are present. However, I have refrained from identifying energies *per se*: it suffices to use temperatures, especially since the CMB, being blackbody, contains a range of photon energies. This in turn means that events tend to occur over a span of time, not instantaneously. Nonetheless, there is always a unique, minimum energy needed to induce any particular microscopic process. Furthermore, any energy can be expressed in terms of an equivalent temperature (which is not a blackbody temperature). You will find a discussion of microscopic energy units and the relationship between energy and its temperature equivalent in endnote 3 of this chapter.

The Ubiquity of the Scale Factor

The scale factor R determines both the size of our homogeneous, isotropic Universe and its Hubble parameter H, as well as the value of the CMB temperature. Because the Universe is expanding, R also measures how much the wavelengths of emitted radiation are stretched and, correspondingly, the amount that frequencies are decreased. In particular, the wavelength λ is proportional to R: $\lambda \propto R$,[4] which in turn implies that the scale factor is related to the redshift parameter z.

The parameter z involves emitted and observed wavelengths, which are the ones of interest: $z = (\lambda_{obs} - \lambda_{emit})/\lambda_{emit}$. If radiation from a receding emitter is observed now, the proportionality $\lambda \propto R$ becomes $\lambda_{obs} \propto R_0$, where R_0 is the current but unknown value of the scale factor. Correspondingly, at the time of emission, the wavelength is λ_{emit} and the scale factor is R_{emit}; they are related by $\lambda_{emit} \propto R_{emit}$. Since λ_{emit} and λ_{obs} occur in the definition of z, it should not be surprising from the preceding proportionalities that R_{emit} and R_0 are the two scale factors that enter the relation between scale factor and redshift.

This relation is given by the following simple formula[5]:

$$R_0/R_{emit} = 1 + z. \tag{7}$$

Because the scale factor R is a measure of the size of the Universe, the value of the ratio R_0/R_{emit} is the amount by which the Universe has increased between the times when the radiation was emitted and observed. The time variation of the ratio can be obtained by solving the relevant equation for R. Using these values of R_0/R_{emit} in Equation (7) leads to the corresponding values of z, some of which are exhibited in the timeline of the Universe that I will discuss in the next section.

Also, in view of the fact that astronomers have created tables of redshifts, such tables can be used to evaluate the corresponding values of the ratio R_0/R_{emit}. This is the reverse of the procedure just outlined. The reversed procedure is analogous to hearing two recordings of an unknown person's voice made at different times and then deducing from them the factor by which the person has grown between the first and second recordings—all without ever having seen him or her!

To illustrate this procedure, I'll choose $z = 6$, which means that the ratio R_0/R_{emit} is equal to 7. Therefore the size of the visible Universe corresponding to a z of 6 was 1/7 of its estimated current size, or approximately 1200 Mpc. This was its diameter at a little less than a billion years after the Big Bang. Since $z \cong 6$ is roughly the upper limit on redshift values that can be extracted from current measurements, they cannot be used to "measure" the size of the Universe at earlier times. Instead, the equation obeyed by R must be solved to yield the value of the ratio R_0/R. Since R_0 is unknown, you may have wondered how the foregoing ratio can be calculated. As it turns out, the solution to the equation for R yields this ratio, so that the unknown value of R_0 is never needed.

The scale factor also determines how long ago the Universe changed from being radiation dominated to being matter dominated. I remarked on this transition on page 142, where the transition time (when d_R and d_M were equal) was stated as approximately 40,000 years after the Big Bang (a more precise value is given a little later in the chapter). The manner in which the scale factor enters the calculation of this number is somewhat complicated to describe, and the analysis, like that yielding Equation (7), is deferred to a chapter note.[6]

Both d_M and d_R become larger as the Universe shrinks, whereas d_Λ does not: it is a constant, independent of R.[a] Consequently, the cosmological constant plays essentially no role in the early Universe. In contrast, the matter and radiation densities grow smaller as the Universe expands, so that as the expansion reaches gigantic amounts, each becomes negligible relative to d_Λ, which is then the only surviving density. This is the reason for the exponential growth of R with time that I mentioned in the preceding chapter, which I have detailed in its second endnote.

A Timeblock

The procedure used in constructing a timeline of the Universe, and especially its early period, has now been identified: one mentally tracks the behavior of the CMB temperature T_{ph} as R decreases and also specifies the densities and the processes induced by the

---

[a]The SDSS analysis examined the possibility that Λ may have changed over time, but as they found no evidence for it, one is safe (at present) in claiming that d_Λ is constant.

increasing energy of the CMB photons. Although this may seem like a straightforward path to follow, you might wonder if indeed it is. Among the questions that may occur to you are the following: is disintegration of structured bodies the only event that can occur as T_{ph} increases; is it possible to go all the way back to time $t = 0$, the instant when the Big Bang happened; and if the answer to the preceding question is no, how far back *can* one go and what are the restrictions involved?

The answers to the first three of the preceding questions are No, No, and roughly 10^{-6} seconds. These responses are interrelated, and I shall explain the reasons for them next, after which I will discuss the timeline.

The No answer to the first question arises partly from the phenomenon known as *pair production*, briefly mentioned at the end of Chapter 4 on the radiation emitted by black holes. Pair production is the creation of a particle–antiparticle pair, and the minimum energy needed to create such a pair is twice the rest mass energy of either of its members, since each has the same mass M. From Einstein's formula, the rest mass energy of one of them is Mc^2, and thus the minimum energy is $2Mc^2$.

Pair production has been observed experimentally. However, it is a mass-dependent process, in that the larger the mass of the particle, the greater the energy needed to create the pair. As an example, consider the electron, whose mass is M_e. It takes a photon of energy at least equal to $2M_ec^2$ to produce an electron–positron pair. On the other hand, because M_e is about 1840 times smaller than the mass of a proton, the minimum photon energy needed to create a proton–antiproton pair is about $2 \times 1840 \times M_ec^2 = 3680M_ec^2$.[b] As a consequence of this mass dependence, the

---

[b]It is an amusing historical footnote that while positrons had been observed in the 1930s (after having been predicted theoretically), physicists nevertheless had uncertainties about the existence of antiprotons. A special accelerator was therefore built to investigate whether the scattering of protons by proton targets could create antiprotons via pair production. In the symbology of Chapter 4, the question to be answered experimentally was whether the process $p + p \rightarrow p + p + p + \bar{p}$ (where $\bar{p}$ represents an antiproton) could occur. For technical reasons,

Universe must become smaller, hotter, and thus younger in order that its population will include increasingly heavier particle–antiparticle pairs.

This last statement implies that by enumerating all the possible particle–antiparticle pairs occurring in the current theoretical framework, it should be possible to go back to the moment just after the Big Bang if not to the instant of it. However, the negative answer to the second question above states that this is impossible. The *timeblock* in the quest to write a complete history of the infant Universe is not a failure to enumerate all possible pairs but rather the absence of a valid theoretical framework that would allow the relevant calculations to be carried out, at least in principle if not in practice.

I examine the timeblock in detail in the next chapter and consider only one of its aspects here: it is the inability, to date, of theorists to formulate a theory that fully accounts for the absence of antiparticles in the Universe. Despite experimental searches for the evidence that would conclusively identify them, none has ever been found. In other words, there is now no naturally occurring antimatter in the Universe. Explaining its absence has become an outstanding problem: why does our Universe consist solely of matter, even though—or despite the fact that—antiparticles can be created by pair production?

There is a hierarchy of conjectural theories that address this problem and that are used to carry out various calculations, although none of them yet satisfies all the conditions required of a valid theory. What most elementary particle theorists and cosmologists accept as reliable is a framework that describes the phenomena that occur once all the antibaryons have been annihilated (by becoming intimate with baryons). The energy at which this occurs corresponds to a time of approximately 10^{-6} seconds after the Big Bang. This time is the one previously identified, and it is

---

the accelerated protons would have needed an energy of about $6M_pc^2$ if the process were to be viable, and the accelerator was designed with this property. The experiment was performed in the early 1950s, antiprotons were detected, and the scientists who conceived and carried out the experiment won the Nobel Prize.

the smallest time discussed in this chapter's timeline. (I am postponing to the next chapter a conjectured timeline of events that may have occurred prior to 10^{-6} seconds, in the very early, infant Universe.)

A Timeline of the Universe

In addition to the times of prerecombination events, recent observations have specified the times when a few postrecombination events occurred, and these are included in the timeline. Bear in mind that the timeline is based on information that was current in 2004 and is therefore subject to future corrections or additions. In this regard, occasional visits to some of the Web sites cited in the Bibliography may yield updates.

The timeline is shown in Figure 25. The first column specifies the particular event, while the time of its occurrence, the corresponding CMB (blackbody) temperature, the redshift parameter z, and the ratio R/R_0 are the quantities stated in the remaining columns. The headings should be self-explanatory, but to make the figure self-contained, the symbols heading the last three columns are identified below the dashed line, along with certain definitions.

As you proceed from the top to the bottom of the timeline, you will encounter earlier and earlier events until you reach antibaryon annihilation, at time of 10^{-6} seconds after the Big Bang. Because the natural progression is from earlier to later times, I shall take the opposite route, thereby examining events in the order of occurrence as the Universe expanded.

All antibaryons were annihilated when the energy in the early Universe was insufficient to create any baryon–antibaryon pairs. This event took place at a temperature of approximately 10^{13} K, when the redshift factor was the gigantic value of 4×10^{12} and the scale factor had decreased by the even greater amount of 3×10^{13}. How big was the visible Universe then? The answer is obtained by dividing its estimated current diameter of 8500 Mpc by 3×10^{13}, which yields a diameter of approximately 28 AU! In terms of distances from the sun, the then-radius of the visible Universe would have had a value between the (average) orbital radii of Saturn and of Uranus. Converting to more familiar terrestrial distances,

Event	Time After Big Bang	T_{ph} (K)	z	R/R_0
Now[a]	13.7×10^9 yr	2.725	0	1
The cosmic jerk	$\cong 5 \times 10^9$ yr	5.34	0.46	0.68
Onset of re-ionization	10^9 yr	15.6	5.73	0.15
Youngest known galaxy	4.7×10^8 yr	25.8	8.5	0.105
First stars form[b]	$\cong 4 \times 10^8$ yr	28.8	9.6	0.094
Recombination (decoupling of matter and radiation) ends[a]	3.79×10^5 yr	3000	1089	9×10^{-4}
Equality of mass and radiation densities[a]	4.1×10^4 yr	9300	3233	3×10^{-4}
Primordial nucleosynthesis begins	$\cong 3$ min	7×10^8	25×10^7	4×10^{-9}
Electron-positron annihilation ends	$\cong 1$ sec	11.6×10^9	4.2×10^9	2×10^{-10}
Quarks form neutrons and protons	$\cong 4 \times 10^{-6}$ sec	4.4×10^{12}	1.6×10^{12}	6×10^{-13}
Annihilation of antibaryons	$\cong 10^{-6}$ sec	10^{13}	4×10^{12}	3×10^{-13}

[a]Time datum from the 2003 Wilkinson Microwave Anisotropy Probe (WMAP) combined results.
[b]Time datum from the WMAP results announced in March 2006.

Primordial nucleosynthesis: early-Universe formation of the nuclei of light atoms—deuterium, helium, and lithium.

Baryons: quarks, neutrons, protons, and their antiparticles (but cosmologists often include e^- and e^+).

Quarks: the fundamental particles that constitute protons and neutrons.

R_0: the (unknown) current value of the universal scale factor R (only the ratio R/R_0 is important).

T_{ph}: temperature of the blackbody spectrum of photons.

z: redshift parameter; $z = (\lambda_{emit} / \lambda_{obs}) - 1$, where λ is the wavelength.

Figure 25. A cosmic timeline: from the early universe to now.

28 AU is approximately equal to 4.2×10^{12} km or 2.64×10^{12} miles.

A little after antibaryon annihilation, at 4×10^{-6} seconds aBB,[c] the expansion had lowered T_{ph} sufficiently that the attraction between the remaining quarks allowed them to coalesce, forming neutrons and protons plus lighter, short-lived particles known as mesons. At this time, the visible Universe had doubled in size and the redshift factor had decreased by nearly a third. And, although

---

[c]aBB, *after the Big Bang*; an abbreviation I will use in the rest of the book.

antibaryons were absent, positrons, the antiparticles to electrons, *were* present: they were being created via pair production.

This state of affairs did not last terribly long: by approximately 1 second aBB, T_{ph} had fallen sufficiently that pair production ceased, resulting in annihilation of all positrons by electrons, in analogy to the previous elimination of the antibaryons. At this time, T_{ph} had decreased by about a factor of 10 and z had fallen by roughly 400, the amount by which the Universe had increased in size. There was, however, too much energy in the hot cosmic soup for neutrons and protons to have initiated primordial nuclesynthesis: the nuclei that were formed were quickly disintegrated by collisions with the enormous number of energetic photons present.

Primordial nucleosynthesis *did* begin about 3 minutes later, when T_{ph} was roughly 7×10^8 K, and it lasted for roughly another 17 minutes, that is, until approximately 20 minutes aBB. The events that occurred during this interval are so crucial for Big Bang cosmology that a separate discussion is warranted. You will find it in the next section, but because primordial nucleosynthesis is a complex subject, I am relegating the discussion of its details to Appendix B, which you may consult at your leisure. I do note, however, that when permanent formation of the very light nuclei began, the diameter of the now visible Universe was about 34 pc and z had decreased from its value at 1 second aBB by a factor of a little more than 200.

Although the contents of the early Universe did not change after the very light nuclei were formed, it was still an active place. The absence of atoms meant that photons were being strongly scattered by the positively charged nuclei and the negatively charged electrons. Furthermore, the Universe continued to expand, so that T_{ph} and z kept on decreasing. Nevertheless, the volume of the Universe was still small enough that it was radiation dominated and remained so until 41,000 years aBB, at which time d_R and d_M became equal. Our now-visible Universe was then about 7.5 Mpc in diameter while the CMB temperature had fallen to approximately 9300 K, somewhat less than twice the sun's surface temperature.

Until the time of recombination, the immediate effect of the transition to matter domination was the change in the time variation of the scale factor: R grew more quickly than it had during the era of radiation domination.[7] Correspondingly, T_{ph} decreased more slowly, but for some time it remained too high to allow for many atoms—mainly hydrogen—to form permanently. In other words, the Universe was still a hot soup of dark matter, light nuclei, electrons, photons, and antineutrinos.

Finally, when the temperature had been reduced to T_{ph} = 3000 K, at z = 1089 and an age of 379,000 years aBB, the number of photons capable of ionizing hydrogen diminished so significantly that recombination occurred. Blackbody photons were then able to travel unimpeded; those now observed as the CMB originated on *the surface of last scattering*. This is a sphere, centered on the earth, whose radius is that of the "Hubble distance at recombination." Its value, based on the combined WMAP parameters, is about 0.2 Mpc.[8]

The recombination time of 379,000 years aBB marks the end of the early Universe. Eventually, gravitational attraction began to concentrate the overdense matter fluctuations into increasingly smaller volumes, whence they could begin the slow process of star and galaxy formation. Analysis based on the latest WMAP results (March 2006) leads to the conclusion that the first stars formed around 400 million years aBB, when the diameter of the now-visible Universe was about 800 Mpc and the CMB temperature was approximately 28.8 K.

When and how the first galaxies came into being is a matter beyond the scope of this book. However, a 2004 Hubble Space Telescope observation has provided an interesting datum in this regard: the most distant, and therefore the youngest, galaxy known at the time of writing is approximately 13.2 billion years old (i.e., it was formed not later than 0.47 billion years aBB). This result can be compared with a 2003 finding from the SDSS team that they may have observed a group of young galaxies, perhaps no more than a billion years old, also an age consistent with stellar formation beginning at 400 million years. More data is needed, and presumably will be forthcoming, to confirm this preliminary result.

Intermediate between the latter two phenomena is another, known as *re-ionization*, that seems to require galaxy and black hole formation. The term *re-ionization* refers to the ionization of neutral hydrogen atoms (formed *via* recombination) by ultraviolet photons that were initially emitted by protogalaxies and accreting black holes, and later by galaxies themselves. Eventually, almost all of the intergalactic hydrogen was ionized into its separate proton and electron constituents. In 2001, the SDSS group reported the first conclusive evidence supporting this scenario. They estimated that re-ionization began at $z = 5.73$ and ended at $z = 5.2$; the time, scale factor, and T_{ph} values that correspond to the onset of re-ionization are listed in Figure 25. The onset time of approximately 1 billion years is consistent with the formation of galaxies noted above. Future observations (on quasar spectra) should yield more information on the nature and z values of the full re-ionization transition.

The last entry in the timeline before "Now" is the *cosmic jerk*, an event that occurred approximately 5 billion years ago. This phrase refers to the transition in the Universe between *deceleration* and *acceleration*. Initially, the effect of matter was to decelerate the expansion, a slowing due to gravitational attraction, similar to the earth's gravity slowing (and ultimately stopping) the flight of a ball thrown upwards. However, the expansion eventually decreased the matter density Ω_M sufficiently that it was overcome by the density Ω_Λ of the cosmological constant. When this happened, the Universe stopped slowing down and started to accelerate. Evidence for the cosmic jerk, expected ever since the acceleration of the Universe had been detected, was first reported in 2003 by the High-Z Supernova Search Team, using information obtained from 16 new type Ia supernovas. They concluded that the cosmic jerk occurred at a redshift of $z = 0.46 \pm 0.13$ (which is listed simply as 0.46 in the timeline), and they drew certain conclusions concerning dark energy (a topic that is an integral part of the next chapter).

Primordial Nucleosynthesis

The wide acceptance of the Big Bang scenario among cosmologists rests on three major pieces of evidence: the cosmic microwave background radiation and its almost perfect blackbody character;

the expansion of the Universe; and the very good agreement between the predicted abundance of the nuclei produced primordially and the measured values, which are obtained from those regions and objects in the sky where the abundance is thought to be primordial. As indicated in Figure 25, primordial nucleosynthesis began at approximately 3 minutes aBB; my estimate is that it ceased at 20 minutes aBB.[9]

I remarked in Chapter 4 that nuclei containing total numbers of neutrons and protons equal to 5 or 8 are unstable. Because of this, just five nuclei are predicted to have formed primordially (they and the reactions that produced them are the subject of Appendix B). However, only four of them, deuterons (d), ^{3}He, alpha particles (α), and ^{7}Li, would have occurred in sufficient abundance to be measurable. The problem is to find sources of them that are conclusively identified as primordial.

The structure of these four nuclei is as follows: ^{7}Li contains 4 neutrons and 3 protons; ^{3}He has 2 protons and 1 neutron; alpha particles (the nuclei of helium atoms) have 2 protons and 2 neutrons; and the deuteron is a neutron–proton bound system. The only ^{3}He found so far are in the solar system and in certain regions of the Galaxy, neither of which is believed to be a suitable primordial source. Hence, only the primordial abundances of ds, αs, and ^{7}Li can be measured.

On the other hand, the theoretical abundance values depend on the ratio of the total number of neutrons and protons to the number of photons, and this ratio, denoted by the lowercase Greek letter η (eta), is uncertain. Theorists deal with this problem by constructing graphs showing how the abundance values change with changes in η. The abundances are actually expressed as ratios of the number of each nucleus to the number of protons. All three light nuclei abundance ratios are in agreement with the measured values, within large error limits, for η in the range 3.4×10^{-10} to 6.9×10^{-10}; the latter range is consistent with result 6.1×10^{-10} obtained from analysis of the CMB anisotropies.

The sources for the measured values of the abundance ratios are certain quasars for the deuterons, clouds of ionized hydrogen in dwarf galaxies for the αs, and certain Population II stars in the Galaxy for ^{7}Li. Errors in the measured results are large but not so large (see Appendix B) to negate the good agreement between theory and experiment that provides the third leg of the tripod of evidence strongly supporting Big Bang cosmology.

Notice, by the way, that because the η values mentioned above are less than 10^{-9}, then the ratio of photons to neutrons or protons in the Universe is more than a billion to one.

9. Conjectures

"To be or not to be . . . ." This famous phrase not only introduces Hamlet's dilemma, it also poses the question a scientific conjecture ultimately must confront. Until that confrontation occurs, the conjecture will be tenable if (1) there is no evidence contradicting it, and (2) it is based either on verified predictions or on a plausible theoretical framework (or both).[a] Sometimes conjectures graduate and become new paradigms, an example of which is general relativity.

The conjectures examined in this chapter were proposed to deal with shortcomings of various kinds, some not previously identified. Inflation, despite its shaky foundation, is the conjecture that has come closest to achieving paradigmatic status, having done so by verification of its predictions. But before considering any of them, I invite you to stroll with me through the garden of elementary particle physics, as it is the background for describing both inflation and how the absence-of-antimatter timeblock is overcome.

Elementary Particles and Their Interactions

Physics is often referred to as the most reductionist of the sciences, since it seeks to explain the complexity observed in the physical world in terms of the simplest possible concepts. Nowhere is this better exemplified than in the quest to understand the structure of matter, where the goal has been to identify the truly elementary constituents common to all matter.

---

[a]"Conjecture" is being used in a very specific way, as indicated by the two conditions stated above.[1] "Speculations," on the other hand, satisfy neither of the requirements noted in the second of these two conditions. Exemplified by "What if . . ." types of statements, they are akin to *ad hoc* proposals. Several will be encountered later in this chapter.

The earliest known proponents of this idea were the ancient Hindus, around 600 BCE, and the ancient Greeks, especially Democritus and Leucippus, ca. 430 BCE [see Teresi (2002)]. For the Greeks, *atom* referred to the fundamental constituent of matter (the word itself comes from the Greek *atomos*, meaning "uncuttable"). From the modern viewpoint, their concept is vague, far more semantic than scientific, and it remained so until the Englishman John Dalton invoked it in the early 1800s. Dalton, who used it in trying to explain the weights of chemical elements, can be considered the "father" of the modern atomic concept.

Although not all 19th century scientists accepted Dalton's hypothesis that atoms were the irreducible elements of matter, this conjecture was undermined over a century ago, first with the 1897 identification of the electron, and later with the discovery of the atomic nucleus. Of these two entities, only the electron is irreducible: it is a structureless, point particle, whereas nuclei are structured objects. The nuclei of all atoms heavier than hydrogen are composed of neutrons and protons, which are collectively referred to as *nucleons*, a nomenclature I will use from now on. Nuclei can thus be reduced to their nucleon substructures.

Despite this reducibility, when considered only on an atomic scale—viz., for lengths of the order 10^{-10} m and energies of the amount needed to ionize hydrogen—nuclei are generally treated theoretically *as if* they were point objects. This treatment is justified because nucleon substructure is generally not observable on the atomic scale. To observe it requires probes whose wavelengths are comparable to nucleon sizes,[2] roughly 10^{-15} m. When electromagnetic radiation is the probe, the photon energies corresponding to these wavelengths are in the nuclear range rather than the atomic one.[3]

Since nuclei are structured objects, it is reasonable to suppose that nucleons might also have a substructure. From comments I made earlier in the book, you know that this supposition is an ingredient of the modern paradigm: it states that nucleons are composed of quarks, which *are* accepted as fundamental entities. In this paradigm—known as the *standard model* of elementary-particle physics—matter at its most fundamental level is divided into three general categories of unstructured particles, called *leptons*, *quarks*, and *bosons*. Appendix C identifies the particles in each of these categories and examines some of their properties.

The 17 elementary particles of the standard model and the fundamental forces governing their interactions and behavior are the ingredients of the theories currently used to describe elementary-particle/high-energy phenomena. Although these theories enjoy varying degrees of success, they are not yet as well founded as general relativity or quantum mechanics; they also contain parameters whose values must be obtained empirically rather than being intrinsic to the formulation. Nonetheless, they and their various extensions have accounted for a wide variety of data, although not all their predictions have been verified. For example, in some extensions of the standard model, protons are predicted to decay, but such decays have not yet been observed, as I shall discuss later.

Because there are only four fundamental forces that govern terrestrial phenomena, then only these four are assumed to be present throughout the Universe. The theories describing elementary particle phenomena incorporate some or all of these forces.

The mightiest of the four is the strong (or nuclear) force, which governs the interactions between quarks, protons and neutrons, and a variety of other strongly interacting particles. For example, it binds nucleons in the nucleus and the quarks that form neutrons and protons. (See Appendix C for the quark substructure of nucleons.)

The next most powerful force is the electromagnetic, which acts between charged particles. It produces the particle-stable quantum states of electrons in atoms.

The third interaction that influences the behavior of the strongly interacting particles is the weak force. It is responsible, among other things, for their decays; for example, the beta decay of the neutron. The weak force is exceedingly short-ranged, essentially requiring contact, whereas the strong force acts over the far "greater" distance of roughly 10^{-15} m.

The fourth force, the weakest of all, is that of gravity. Because it is so puny in the microscopic world (recall from Chapter 3 that for two protons it is about a factor of 10^{36} less powerful than the electromagnetic force), gravity is excluded from theoretical frameworks other than string theory and loop quantum gravity, which I will consider later. Here I will focus on theories that impinge on the timeblock and inflation.

Symmetry Properties of Elementary Particle Theories

Symmetry concepts are a signal feature of modern theories in physics, as I have commented several times. The two encountered in Chapter 6, homogeneity and isotropy, are properties that matter will enjoy under certain circumstances, in which case they will be manifested when either position changes along a fixed direction—known as a *spatial translation*—or direction changes at a fixed position—known as a *rotation*. These changes are effected by carrying out the *symmetry operations* of translation or rotation.

Symmetry operations play a critical role in formulating elementary-particle theories, in particular the three operations known as *charge conjugation*, *spatial reflection*, and *time reversal*. They are symbolized, respectively, by the capital letters C, P, and T, where P denotes *parity*. Each produces a different effect when it acts.

Charge conjugation is the operation that replaces all the particles in the system by their corresponding antiparticles while simultaneously replacing all the antiparticles by their corresponding particles. (One way of illustrating how this operation would be carried out is to imagine changing the electric charge of each particle in the system by a charge of the same magnitude but the opposite sign.)

Spatial reflection is the operation that replaces each of the three spatial directions—for example, height, length, and width—by the opposite directions. In equations involving spatial directions, this operation is effected by changing the sign, from positive to negative, of the lengths along each direction. To understand what this means, stand a given distance in front of a full-length mirror. Your image will appear to be at the same distance *behind* the reflecting glass as you are in front of it. If your distance *to the* mirror is considered to be *positive*, then your image's distance *behind* it is *negative*, and *vice versa*.

The operation involved in *time reversal* is the theoretical one of causing time to run backwards. A simulation of time reversal occurs every time a film is run in reverse. Perhaps you mentally engaged in this operation when going through the procedures involved in setting up the timeline of the preceding chapter. Time is usually represented in equations by the symbol t. The simplest means of effecting time reversal is by *changing the sign* of this symbol from positive to negative, in other words, of replacing t with $-t$. It is the analogue of the operation that produces spatial reflection.

For about 10 years after the end of the Second World War, it was conventional wisdom that applying each of these symmetry operations to any microscopic physical system would not change the outcome of experiments.[4] In particular, it was taken for granted that if two experiments were carried out, one before and the other after application of P, then the *same* result would be obtained. This is referred to as *conservation of parity*. Weak interaction systems were not only believed to *conserve* parity (P), but physicists thought that experiments testing conservation of parity *had been* performed.

This complacency was shattered in 1956 by two Chinese-American theorists, Tsung Dao Lee and Chen Ning Yang. They pointed out that measurements had never been carried out to test spatial reflection symmetry for weak interactions, and they also proposed experiments that could reveal if weak interactions *were* invariant (see note 4) with respect to P. In less than a year, just such an experiment was performed at the U.S. National Bureau of Standards (now known as the National Institute of Standards and Technology).

The result, which took the world of physics by storm and shook it up, was that spatial reflection is not a valid symmetry operation for weak-interaction systems, the insight for which earned Lee and Yang the Nobel Prize. This particular lack of invariance means that parity is *not* conserved for weak-interaction systems. Furthermore, not only is conservation of parity violated by the weak interactions, they don't conserve charge conjugation symmetry (C) either.

The failure of the weak interactions to conserve these two symmetries led to a paradigm shift, one consequence of which was a recasting of the theory of weak interactions into a form that was explicitly not invariant under the action of P or C. And, because each of P and C was not a valid symmetry operation for the weak interaction, a new question arose: would it be invariant under the combination of the two symmetry operations (either one followed by the other—CP or PC)? The answer was No, as discovered in experiments carried out in 1964 by a group at Princeton University led by the Americans James Cronin and Val Fitch. They found that the weak-interaction decays of a certain meson were not invariant under the combined symmetry operations of C and P; that is, non-invariance with respect to this combination was established.[5]

Cronin and Fitch won the Nobel Prize for this research. Their discovery was immediately followed by the realization that time-reversal symmetry is also violated. That is, none of C, P, CP or PC, or T is a valid symmetry property of the weak interaction.[6] Once again, there was a recasting of weak-interaction theory so that it took a form in which none of the preceding symmetry properties were valid operations for it. This was done in an empirical way, rather than from first principles, since there was then no theory in which time-reversal non-invariance is a natural element. The reformulation eventually provided a mechanism that allows for a baryon–antibaryon asymmetry in the very early Universe—a hugely important consequence.

The next major advance was unifying the theories of the weak and electromagnetic interactions into a single framework known as the *electroweak* theory. Unification was a joint accomplishment, achieved through the efforts of Abdus Salam, a Pakistani physicist, and two Americans, Sheldon Glashow and Steven Weinberg; the three later shared the Nobel Prize for this work. They realized that the weak and electromagnetic forces could be recast as a single force as long as the energy was above a certain threshold, one corresponding to a temperature of approximately 3.5×10^{15} K. Particles with this energy have been produced by accelerators such as the Tevatron at Fermi National Accelerator Lab in Illinois. For energies below the threshold, the electroweak theory splits into the two separate portions from which it was created.

Electroweak theory is one component of the standard model of elementary particle physics. Its other component is *quantum chromodynamics*, the strong-interaction theory governing the behavior of quarks. The standard model has had many successes predicting and correlating experimental data; hence its name. Among the successes is the prediction of the size of the CP symmetry violation in the very rarely occurring and hard-to-detect weak decay of the "B" meson, a prediction that was triumphantly verified in 2002. It was just the second instance of a CP-violating decay.

Despite its successes, the standard model cannot be the last word in elementary particle theory, because (1) it contains many parameters whose values are unspecified and therefore can only be obtained by fitting data; (2) its left–right (parity-violating) asymmetry is artifactual, put into the theory "by hand" rather than being a fundamental element; and (3) the arranging of its six quarks and its six leptons into "families" of three pairs each lacks a fundamental basis and is therefore empirical as well.

Theorists have tried to provide a firmer theoretical basis for the standard model by creating a variety of frameworks called grand unified theories (the acronym, of course, is GUTs). These theories are called "grand" because they combine the already unified electroweak theory with quantum chromodynamics to form overall unified theories. In them, quarks and leptons are "unified" in a way that makes them interchangeable. This particular unification occurs at temperatures (energies) greater than about 10^{27} K (!!), the threshold at which the strong and the electroweak forces become equal.

GUTs have entered my narrative because they solve several outstanding problems. For instance, they allow neutrinos to have mass (needed to solve the solar neutrino discrepancy; Chapter 5), and they establish the equality (in magnitude) of the electron and proton charges, an equality not previously understood. (The fact that the charges of the electron and proton are equal in magnitude is regarded by physicists as a phenomenon to be explained and not simply as an accident of nature. That is, like countless other phenomena, it should be understandable in a reductionist way, in analogy to quantum theory explaining the Periodic Table of the elements or the scattering of light by particles in the atmosphere explaining the blue color of a cloud-less sky.)

From the perspective of this book, however, the most important problem solved by GUTs is their providing a mechanism that explains the observed absence of antibaryons in the Universe. It is a consequence of quarks being able to undergo a normally forbidden, CP-violating decay, an event that occurs because quarks and leptons can be interchanged in grand unified theories.

Although GUTs seem to solve the antibaryon problem, in fact they do not quite do so. The reason is that their CP symmetry violation is that of the electroweak theory, which is put in by hand, rather than being a first-principles result. There are other drawbacks to GUTs as well. A major one is their prediction that protons are unstable: because of the lepton–quark interchangeability, protons are predicted to decay. One prediction for the proton's lifetime is about 10^{30} years.

You should not be lulled by this enormous number into thinking that it renders proton decay unobservable. "All" that is needed are a detector sensitive only to the decay products and a mass containing more than 10^{30} protons. That is, if it takes 10^{30} years for a single proton to decay, then among 10^{30} of them one should decay each year. Since the inference from Table 2 is that an adult person contains more than 10^{28} protons, obtaining the requisite mass is not a problem. Indeed, proton-decay experiments *have* been carried out, but the decay has never been observed. Its absence, coupled with the amount of material used as the source of possibly decaying protons, has led to an empirical value for the minimum proton lifetime. It is approximately 10^{33} years, a number too large to be consistent with any of the GUTs. Hence, the failure to detect the decay is another reason that GUTs are not quite the final framework but instead are waypoints on the path toward more plausible theories.[7]

Inflation: Three Puzzles and Their Solution

The groundwork is now in place for me to finally examine two important conjectures: inflation and a mechanism for dealing with the timeblock. I will consider inflation first, partly because it was introduced to solve a set of puzzles, one of which is connected with GUTs.

The new GUTs problem is the failure to verify the prediction concerning an entity known as a *monopole*, which is an isolated north or south magnetic pole. It is an analogue of a positive or a negative charge. GUTs require that copious numbers of monopoles should exist; none, however, have ever been observed. They cannot be made by cutting a bar magnet: no matter how many subdivisions are made by slicing it into smaller and smaller pieces, each little piece always has both a north pole and a south pole. Nonetheless, while electromagnetic theory can be modified in a way that *allows* monopoles to exist, GUTs *demand* them. In view of this, the question was where are the monopoles?

The absence of monopoles was not the only puzzle. Two others that concerned theorists are related to Big Bang dynamics. One involves the strong likelihood that our Universe is flat. Recall that in a flat Universe the curvature parameter is zero, which from Equation (6) requires that the total density $\Omega_0 = 1$. But, in order that the total density be unity now (to within experimental uncertainty), it has been shown that at times very close to the instant of the Big Bang, it must have had the value $\Omega_0 = 1 \pm 10^{-N}$, where $N = 52$ or 56 (!!), the particular value depending on the time aBB that is used in the calculation. The number 1 ± 10^{-52} differs from unity by 1 in the 52nd decimal place, and this is where the puzzle lies, for such a number implies that the Universe was incredibly—almost unbelievably—fine tuned to flatness at an instant following the Big Bang.

Although such fine tuning might be possible, a physical mechanism allowing it to occur has not been found; hence, it is a cosmological puzzle awaiting solution.[b] Finding the key to the puzzle is analogous to finding an explanation for the equality in magnitude of the electron and proton charges.

The remaining puzzle is known as the *horizon problem*. It arises because the microwave background should not be (nearly) isotropic, much less blackbody. Why? The answer is that prior to

---

[b]One way, of course, to account for the fine tuning is via divine intention, a nonscientific explanation that lies outside the purview of this book.

the time of recombination, the early Universe could not have been in an equilibrium state and therefore could not have had a temperature. The reasoning underlying this conclusion is that for the baryons to have achieved equilibrium, there had to have been sufficient contact between them and the photons. However, the Universe was expanding too rapidly for radiation to have interpenetrated all of the baryonic matter. The absence of such photon penetration would have prevented the occurrence of equilibrium and without it, a temperature would be impossible, thus denying a blackbody character to the photons. And that means: no CMB!

The inability to achieve equilibrium is called the horizon problem because the greatest distance photons could have traveled prior to recombination is the *particle horizon*, a distance that is always *less* than the diameter of the Universe. At the time of recombination (Figure 25), the visible Universe was about 28 million light years in diameter, whereas the particle horizon distance is estimated to have been somewhat less than 1 million light years. The ratio of the two distances is about 0.04, a number that becomes even smaller as one goes farther back in time. Reaching equilibrium is therefore always precluded. But because the CMB exists, equilibrium *was* achieved, and the question is, how was it accomplished?

The answer to this question is by means of inflation, which just happens to solve the other two puzzles as well. The solution was published in 1981 by Alan Guth, a young postdoctoral fellow then searching for a permanent faculty position. Although inflationary ideas were in the air at that time, he is credited with providing the answers. The premise underlying inflation—which involved concepts from GUTs—is that at an extremely early time, say 10^{-36} sec aBB, the then very tiny Universe experienced a period of extraordinarily rapid growth. It lasted until approximately 10^{-34} sec aBB, increased R by a factor in the range 10^{28} to 10^{30}, and occurred when grand unification or a similar synthesis governed the Universe.

The vast increase in R resolves all three puzzles. First, while monopoles *were* present, the huge expansion swept them all so far away from us and each other that the probability of observing one has become negligible. Second, inflation causes the Universe to grow so large that its inherent curvature cannot be ascertained: it is so huge that it appears flat, just as the earth does in the immediate vicinity of your location. Hence, $\Omega_0 = 1$ without a need for fine tuning. Finally, prior to inflation all parts of the Universe *were* in contact, equilibrium *was* achieved, and the radiation *did* attain a blackbody character. And then, when inflation occurred, the equilibrium condition of its plasma was not disturbed, and the CMB remained blackbody.

What a neat way to solve these puzzles! You can now see why this conjecture seems to be well along the road to achieving paradigmatic status. For more details than I have provided, plus comments about the sociology of a physics job hunt in a tight market, you should consult Guth (1998).

And there is more! In addition to solving the foregoing problems, inflation predicts that the CMB anisotropies should be polarized and their power spectra should display peaks and valleys that should damp out (diminish), predictions that have been verified, as I noted in Chapter 7. Furthermore, the power spectra are consistent with two other predictions concerning technical characteristics of the anisotropies, viz., that they are *Gaussian* and *scale free* (although the meaning of these terms is not germane to this discussion). Yet another verified prediction is that a typical overdense region at about 500,000 years aBB should have grown via expansion to a size of about 40 million ly, which is just about the diameter of a cluster of galaxies.

Despite these successes, inflation is a theory still not quite at the paradigm level. It is a framework that grew out of ideas indigenous to grand unified theories and makes essential use of a GUTs ingredient known as a *scalar field*.[c] The behavior of this entity is

---

[c] The field concept is one of the most basic in modern physics. Fields may be thought of as disturbances in the fabric of spacetime, typically generated by particles. For example, electromagnetic fields are generated by charged particles; gravitational ones by masses. Fields propagate at finite speeds (the speed of light for electromagnetic and gravitational fields) and are the carriers, so to speak, of the fundamental forces. Their permeation of the space surrounding their source(s) eliminates the unsatisfactory concept of forces that act over a distance, a concept with which even Newton, who introduced it, was unhappy. Since forces are more familiar to readers without a physics background, I have used forces rather than fields as the primary expression of the interactions between elementary particles.

crucial to the starting and ending of the period of inflation, and initially these aspects were problematic. Although this situation has improved through the work of many researchers, the details of the scalar field—it is called an *inflaton field*—are still not on a completely firm footing.

Inflatons are the particles associated with the inflaton field, in analogy to photons being the massless particles associated with the field underlying electromagnetic radiation. While observation of inflatons would go a long way toward further validation of the inflation conjecture, it is the absence of a solid theoretical foundation that is the main obstacle, as I remarked previously. Despite this absence, inflation is widely regarded as *the* mechanism that solves the monopole, flatness, and horizon puzzles, and attempts to put it on a firm basis will likely remain a forefront research activity for some time.

The Absence of Antimatter

The timeblock introduced in the preceding chapter represents the absence of a firmly established theoretical framework that explains the absence of antimatter in the Universe. Because of this, the annihilation of antibaryons at 10^{-6} seconds aBB was the earliest event that could be identified without resorting to conjectures.

As an introduction to this topic, I will first consider the claim itself, namely, that antimatter is not present in the Universe.[8] This claim is based on the terrestrial result that whenever an antiparticle meets up with its particle partner, the two annihilate, creating high-energy radiation in the form of a pair of gamma rays. Observation of this signature radiation therefore establishes that particle–antiparticle annihilation has occurred. Extrapolating to the Universe, the scenario is that *if* there are antiparticles present, some should eventually meet up with their particle partners, whereupon they would annihilate and produce high-energy gamma rays. Many experimental searches for this radiation have been made; none has been found. The absence of such gamma rays is therefore accepted as evidence of the absence of antimatter.

There are three different ways to understand this situation. It may be that there *are* antiparticles in the Universe, but they just haven't been found, so experimenters should keep on looking for them. However, this is not the conventional wisdom, which assumes that sufficient searching has occurred. Or, one may simply accept this absence as an aspect of the Big Bang: simply more particles than antiparticles were created and then, after all the antiparticle annihilations took place, enough matter was left over to populate the Universe with the material objects it contains. There are two grounds on which this scenario has been found unappealing. First, it is unsatisfying to those who require understanding by means of a detailed mechanism. Second, it violates what has become an inherent expectation among cosmologists and elementary particle theorists that symmetry should be a characteristic of the Universe.

The meaning of the latter statement is that if in addition to radiation, the Big Bang created both particles and antiparticles, it should have produced them in equal amounts. Furthermore, if only radiation was present an instant after the Big Bang, then equal numbers of particles and antiparticles had to have been produced subsequently (via pair creation). One of two events must have occurred next. Either the expansion would have separated some antiparticles from their particle partners, precluding annihilation of all pairs but leading eventually to occasional annihilations somewhere in the Universe (however, their signature radiation has not been detected) or *all* pairs would have been annihilated, thereby leading to a Universe containing only radiation, which is certainly not the case.

Since neither of the preceding ways of understanding the antimatter problem has proved to be satisfactory, a third one is obviously called for. Needed is a mechanism that will eliminate all of the antiparticles but not all of the particles. Using the primordial nucleosynthesis result that the ratio of baryons to photons in the early Universe is approximately 6×10^{-10}, cosmologists have determined that an excess of baryons to antibaryons of about one per billion would have populated the Universe with its observed amount of gas, dust, stars, and galaxies. In other words, a mechanism is required that will lead to roughly $10^9 + 1$ baryons for every 10^9 antibaryons prior to pair annihilation.

Such mechanisms have, of course, been put forward. They are based on GUTs or on generalized versions of them that incorporate a framework known as *supersymmetry*. Supersymmetry, acronymically referred to as SUSY, is a theoretical framework in which the leptons and baryons of Appendix C have so-called supersymmetric boson partners, and *vice versa* (see note 7). It is the SUSY partners that mediate the weak decays, which in turn lead to an asymmetric creation of particles and antiparticles. The former are present in a slight excess over the latter, as required, thus leading to a Universe in which antimatter is absent.

Because SUSY-type decays produce an excess of baryons over antibaryons, the *difference* between the number of baryons and the number of antibaryons before the decay is not the same as it is afterwards. Such a change is in contrast with the situation in both strong and electromagnetic processes, where the difference *is* the same before and after. An analogous situation can also arise in the case of the lepton–antilepton difference. A further point is that decays that do not conserve the number of baryons or leptons can only occur under nonequilibrium conditions, in contrast with the requirement for producing blackbody radiation. These are often fascinating but complex details to learn about, and I refer interested readers to the references cited in endnote 7 to this chapter.

The procedures just outlined do lead to a preponderance of matter, but because they are based on theoretical schemes that are not yet fundamental—among other features they contain adjustable parameters whose values are determined empirically—they must be regarded as conjectural. Nevertheless, they are successful, and I shall use them to slide past the timeblock at 10^{-6} seconds aBB and to create a conjectured timeline for the very early Universe.

The conjectured times and events to be featured in the timeline are the onset of baryon asymmetry (more quarks than antiquarks, and also more leptons than antileptons), the earlier periods when inflation and GUTs-type theories governed the Universe, and finally, the earliest era of them all, during which quantum gravity effects are thought to be relevant.

Going backward in time from quark–antiquark annihilation leads, after a "huge" time interval has been covered, to approximately 10^{-34} seconds aBB,[9] the conjectured onset of baryon asymmetry. The temperature was then roughly 10^{27} K, leading to a redshift parameter $z = 3 \times 10^{27}$ (!) and a scale factor ratio of $R/R_0 = 4 \times 10^{-28}$. This ratio implies that the diameter of the visible Universe was about 10 cm!!

The latter size, obtained by a straightforward reversal of the expansion, shows you how concentrated the Universe was moments after the Big Bang. However, this tiny diameter refers only to the visible portion: the entire Universe may well have been larger, even much larger at that time. Absent definite information on this point, I shall continue to concentrate on the visible portion. Doing so leads to an understanding of how incredibly small the visible Universe was when inflation began.

Inflation, estimated to have begun at roughly 10^{-36} seconds aBB and ended at 10^{-34} seconds aBB, occurred too rapidly for the initial temperature to have diminished very much, dropping only by a factor of 10 from the value of 10^{28} K at the onset. More significantly, inflation increased the size of the Universe by a factor of approximately 10^{28}! In other words, R/R_0 was about 10^{-56} when inflation began, and thus the visible Universe occupied a sphere so small—10^{-30} m in diameter—that its size is beyond easy comprehension. If such a size leads you to question the validity of an extrapolation based on the scale factor of general relativity, then you are in very good company. It is probably the most important question one can ask about the physics of the very early Universe.

The GUTs era is estimated to have begun, for reasons cited below, at a time of approximately 10^{-43} seconds aBB, when the temperature of the Universe is thought to have had a value of roughly 10^{32} K. Its now-visible portion is believed to have been no smaller then than approximately 10^{-35} m. The time of 10^{-43} seconds aBB, and the corresponding length scale of 10^{-35} m, are numerical values that characterize the end of the so-called *Planck era*, named for the physicist who solved the blackbody radiation problem (discussed in Chapter 3).

Defined as the period of time prior to 10^{-43} seconds aBB, the Planck era is the one in which quantum phenomena must be taken into account when describing the behavior of the infant Universe. To do so properly requires a quantum theory of gravity, one for which the scale-factor extrapolation need not be valid. So far, of course, the only theory of gravity that has entered the description of the Universe has been general relativity, a pre-quantum framework from which quantum-type entities are absent.

Quantum theory governs the behavior of microscopic systems (it underlies the standard model, electroweak unification, GUTs, etc.), and its signature ingredient is a quantity known as Planck's constant, denoted here by the symbol $\hbar$. Expressed in units of energy $\times$ time, $\hbar$ is so tiny that quantum theory plays no role in the description of everyday phenomena.[10]

In contrast, Planck's constant plays a fundamental role in the description of microscopic phenomena. It occurs in the formula for each measurable quantity that can be calculated using quantum theory, in particular all energies, lengths, and times. Examples are the energies of atoms, molecules, and nuclei; the lengths characterizing electric and magnetic properties of the preceding objects; and times such as the lifetimes of decaying states.[11] The preceding systems are, of course, microscopic; only a few macroscopic systems display quantum effects, two examples being white dwarfs and neutron stars.

In view of the fundamental role that $\hbar$ plays in determining microscopic times, lengths, and energies, you may feel a sense of consistency—or perhaps a sense of the unity of the Universe—to learn that Planck's constant, in conjunction with the speed of light and Newton's constant of gravity, can be used to specify a time, a length, and an energy. Taken together, they characterize the Planck era.[12] The first two have been stated already; the energy, expressed as a temperature, has the approximate value 10^{32} K, which is the approximate temperature when the GUTs epoch began (and the Planck era ended). In that epoch, a GUTs-type framework coupled with general relativity may well be sufficient to describe the behavior of the Universe.

The Planck temperature (denoted T_{Pl}) is presumed to be the *minimum* one at which quantum effects should become important in a quantum theory of gravity. Such a theory might turn out to be the hypothetical but long-desired "theory of everything" (ToE) that elementary particle physicists have dreamed of for years

[Weinberg (1994)]. If a ToE is ever achieved, it is expected to reduce to general relativity plus a GUTs-type framework at an appropriate energy. By its very definition, the ToE would explain the values of all particle masses, the reason why there are only three families each of quarks and leptons, the *modus operandi* of inflation, symmetry violation as a natural ingredient, etc, etc. These hypothesized yet necessary features should make clear why it is denoted a "theory of everything". To date, there are two candidates that might turn into the ToE, one of which—string theory—has been highly publicized. I shall consider each of them in the context of some conjectures put forward to explain the acceleration of the Universe.

However, it is unnecessary to know any more about the ToE than the Planck time, the Planck temperature, and the Planck length to create a conjectured timeline for the very early Universe. It is illustrated in Figure 26 and summarizes the main themes of the preceding discussion. A value for z is stated only for the onset of baryon–antibaryon asymmetry, since the gigantic expansion of inflation precludes meaningful values for it at earlier times. Furthermore, no value is given for R/R_0 during the Planck era, since

Event	Time After Big Bang	T_{ph} (K)	z	R/R_0
Baryon/antibaryon asymmetry ($\cong$ 1 extra baryon per 10^9)	$\cong 10^{-34}$ sec	10^{27}	3×10^{27}	4×10^{-28}
Inflation (?) begins – ends	$\cong 10^{-36}$ to 10^{-34} sec	10^{28}		10^{-56} to 10^{-28}
Grand unified type of theory?	$\cong 10^{-43}$ to 10^{-36} sec	10^{32} to 10^{28}		Less than 10^{-56}
Planck era: For times prior to t_{Pl}, a quantum gravity theory is needed	$\cong 10^{-43}$ sec	10^{32}		[Length$_{Pl}$ $\cong 10^{-35}$ m]

Baryons: quarks, neutrons, protons, and their antiparticles (but cosmologists often include e^- and e^+).

Quarks: the fundamental particles of which protons and neutrons are comprised.

R_0: the current (unknown) value of the universal scale factor R (only the ratio R/R_0 is important).

T_{ph}: temperature of the blackbody spectrum of photons.

z: redshift parameter, $z = (\lambda_{emit} / \lambda_{obs}) - 1$, where λ is the wavelength.

Pl: abbreviation for Planck.

NB: The numbers above the dashed line are obtained assuming that now the relative densities are $\Omega_0 = 1$, $\Omega_M = 0.3$, $\Omega_\Lambda = 0.7$, where $0 = $ total, $M = $ mass, and Λ is the cosmological constant.

Figure 26. A conjectured, very early, cosmic timeline.

general relativity is not expected to be a valid theory of gravity then, as per my comments above.

Dark Matter

The composition of dark matter has been a mystery almost from the moment its existence was deduced. A number of possibilities have been proposed, and I will review four of them in this section. Two are exotic in character and are thus truly conjectural. I start with the nonexotic possibilities.

"Nonexotic" in this case simply means that the proposed candidates are well-known entities, at least in the context of astronomy and cosmology. The first of them is the set of objects known as brown dwarfs.[13] Briefly mentioned in Chapter 2 as "failed stars," they are a category of star whose masses are too low to allow nuclear burning to occur. The key phrase here is "category of star." Only stars are included in this category because of the mode of formation, which is a collapse under gravity of a cloud of gas and dust. They "shine" by emitting photons whose energy is supplied by gravitational contraction. In contrast, planets are thought to have formed by coalescence of small rocky and/or icy bodies and are believed to have a maximum mass not much greater than that of Jupiter.

Until 1994, brown dwarfs were a theoretical construct, expected to exist but not yet observed. The first one was discovered in that year, and since then many more have been detected. Their masses are believed to lie between 10 and 70 times the mass of Jupiter, while their luminosities are thought to be at most $10^{-4} \times L_{Sun}$, features that make their detection very difficult.

In 2000, astronomers found two plentiful sources of young brown dwarfs ("young" because their luminosities were about equal to the figure listed above, whereas older ones at the same distance would be fainter). They concluded that brown dwarfs are probably as numerous as ordinary stars. It is their low luminosities that make them a possible dark-matter candidate. However, were they as numerous as just cited, and if almost all of them were to have the maximum mass, their contribution to dark matter could only be a few percent. There would need to be far many more than are currently thought to exist for them to be the primary component of dark matter.

The brown dwarf star is a nonexotic baryonic candidate. The other nonexotic candidate is the triplet of neutrinos, which are leptonic, not baryonic. Neutrinos interact extremely weakly with baryons and are now known to have small masses. Do they occur in sufficient numbers to be a major dark-matter candidate? Apparently not. The reason is the recent upper limit on neutrino masses reported by the Sloan Digital Sky Survey (page 169). That upper limit is much less than the ones that Peacock (1999) and Raine and Thomas (2001) require in order to make neutrinos a significant component of dark matter. At best, then, their contribution cannot be significant. Other candidates are therefore needed, and I turn next to the two exotic possibilities. Both are conjectural, neither has been observed, and experimental searches for each are being carried out.

The two exotic particles belong to the category of WIMPS, an acronym for weakly interacting massive particles. Denoted the *axion* and the *neutralino*, they are the current favorites among the cosmology community. The first is a heretofore uncited particle of the standard model, while the second is the lightest of the SUSY particles.

Axions are hypothetical particles that arise when CP-violating processes that involve baryons are eliminated from the standard model. Such processes were allowed by the mathematical structure of quantum chromodynamics (QCD). Because they are not observed in practice, QCD was altered by a mechanism that prevents their occurrence. The axion is a consequence of this preventive action.

Theoretical analysis suggests that a mass of roughly a millionth of an electron's mass would be sufficient for axions to constitute all of dark matter. If axions do exist, their mass could differ from this value, and unfortunately, the axion mass is an unknown parameter in the theory. Individual experiments that search for axions are limited in the range of masses to which the equipment is sensitive, so that possible detection of these weakly interacting, hypothetical particles could take quite a long time.

Neutralinos are one of the hypothetical "symmetry partners" in the SUSY extensions of the standard model that I introduced on page 206. Since theorists find SUSY very attractive, its framework and particles are taken seriously. Neutralinos are the lightest of the SUSY particles, but as with axions, their mass is an unknown parameter. Estimates range from about 100 to 600 times the proton mass, making them far heavier than axions, and experiments looking for neutralinos have been set up to search within narrow portions of this range. Detection is difficult, because neutralinos interact weakly with baryonic matter, the stuff of which detectors are made.

There are two broad categories of dark-matter candidates: *hot* and *cold*, the former referring to particles whose speeds are equal or close to the speed of light, whereas the latter move much more slowly. Axions and neutralinos are examples of cold dark matter, and it is for such particles that current searches are being carried out. A number of groups are either setting up or actively performing such searches, which are among the "hottest" experiments of the current era. In spring 2004, the results of two searches carried out underground were inconclusive, with one group asserting positive evidence for WIMPS, and the other finding no evidence. If you want to keep up to date in this exciting field, I suggest accessing one of the appropriate Web sites listed in the Bibliography or simply typing "dark matter" into your favorite search engine.

Acceleration of the Universe

Most of the words, phrases, and acronyms that relate to the conjectures I have dealt with are well out of the ordinary: nucleon, parity, charge conjugation, time reversal, electroweak, quantum chromodynamics, scalar fields, inflatons, axions, neutralinos, GUTs, SUSY. Clustered together this way they may seem like ingredients in a recipe for an imaginary stew. Nevertheless, such words are quite consistent with the extraordinary nature of the conjectures themselves. In contrast, while the phrase "acceleration of the Universe" involves the common word *accelerate*, the conjectured solutions to the problem posed by the acceleration involve concepts no less extraordinary than those already encountered. The one I shall start with is Λ, Einstein's "greatest blunder."

(Because you've seen it so much already, it may seem ordinary, not extraordinary.)

So far, the acceleration of the Universe has been attributed to the presence of the cosmological constant. Values for its relative density have been deduced from supernova and CMB anisotropy data. However, as I remarked in Chapter 7, the value of Ω_Λ is far smaller than certain theoretical estimates indicate it should be. These estimates are based on two assumptions.

The first is that Λ represents a type of energy associated with the vacuum. Far from indicating empty space, however, the physics vacuum is seething with particle–antiparticle pairs that rapidly wink into and out of existence. Known as *virtual pairs*, the energy needed to create them is "borrowed" from empty space. By means of Heisenberg's uncertainty principle, quantum theory allows such borrowing to occur, but only for miniscule periods of time.[14] Since the energy is borrowed, it must be given back, whence the pair so created vanishes without producing photons (none were there to begin with, and therefore none can be created afterwards—which is why the pairs are termed "virtual"). Virtual pairs are continually popping into and out of existence, thereby giving rise to a nonempty vacuum.[15]

The second assumption used in estimating Ω_Λ is composed of two parts. First, Λ is identified *as*, and not just *associated with*, vacuum energy. Second, the vacuum energy is assumed to be the Planck energy mentioned above. This latter choice reflects the belief that Λ came into existence during the Planck era. On transforming the Planck energy into a relative density (division of the equivalent mass by the critical density d_c), the estimated value of Ω_Λ is approximately 10^{123}, a number to be compared with the measured (combined WMAP) value of 0.73!!! As Rich (2001) remarks, the estimated value "is perhaps the worst guess in the history of physics."

With respect to the current understanding of cosmology, the value $\Omega_\Lambda \cong 10^{123}$ is totally absurd, since the acceleration associated with it would never have allowed the Universe to form. The colossal discrepancy between the predicted and measured values obviously raises questions, one of which is the following: is any of the input leading to the predicted value correct? At present, this query appears not to have a reliable answer, although proposals have been put forward. In some, the standard concept of the vacuum is modified. For instance, in one conjecture a certain black hole feature denoted a *hologram* changes the standard elementary particle vacuum. Here, "hologram" is used as an analogue to holograms in optics, wherein a three-dimensional object is reproduced by a two-dimensional image. Although the resulting change in the vacuum produced by the hologram idea could allow Ω_Λ to have its observed value, there have been no experiments proposed yet that would test this conjecture.

While the cosmological constant was the original explanation for the acceleration of the Universe, another is in terms of a more general concept called *dark energy*, a term used in both popular and scientific treatments. Dark energy is an ingredient in a pair of questions raised by the preceding discrepancy: can the acceleration of the Universe be a consequence of a dark-energy mechanism other than vacuum energy (which many cosmologists believe Λ to be) or does it arise from a totally different mechanism?

A variety of conjectures have been put forward as answers to these questions, and I shall consider four of them. They involve the following entities: a hypothetical quantum field denoted *quintessence* (quantum fields are addressed in footnote c); the axion; string theory; and loop quantum gravity (the latter two are replacements for Einstein's theory of gravity).

Quintessence is a dark-energy quantum field that presumably permeates the Universe. It may be a scalar field like the one postulated in connection with inflation. Unlike vacuum energy, however, and depending on the model used to carry out the calculations, the relative density of the quintessence field in the early Universe may not be too different than the current value of Ω_Λ; it would slowly decrease with time, doing so in such a way as to be larger now (and also at the time of the cosmic jerk) than Ω_{M0}. This would not only lead to the accceleration, it would also eliminate the problem of an Ω_Λ whose value is 10^{123} in the very early Universe. That is, the need for an almost perfect (and incredible) cancellation of the factor of 10^{123} would no longer exist.

Although quintessence could produce the observed acceleration, verification of this conjecture is possible only if an effect unique to it can be observed. Several have been proposed. One uses the fact that quintessence, like the vacuum energy that Λ represents, gives rise to a negative pressure. The ratio of this pressure to the density differs for it and Λ; it is possible that higher-accuracy supernova measurements could determine the ratio with the precision needed to distinguish between the two mechanisms.

Effects peculiar to quintessence could also show up in the CMB. Detection, however, will require improved precision, and in any case it may be difficult to separate them from other influences. (For more about quintessence, see the monograph by Krauss cited in the Bibliography. You might also try typing *quintessence* into a Web browser and go from there.)

In contrast with the other conjectures, the one involving axions claims that the Universe is not actually accelerating, it only *appears* to be. The experimental evidence for acceleration is the fact that distant supernovas are *dimmer* than is consistent with the distances measured to them; that is, their apparent luminosity is less than expected. This normally would mean that they are further away than thought. But if they are more distant, then their expansion speeds must be greater than predicted from Hubble's law, and it is this result that requires the Universe to be accelerating.

In the axion explanation, however, the supernovas *are* at the proper distances, but *fewer* photons are being detected, which leads directly to a decrease in the apparent luminosity. Why is the number of photons diminished? The conjectured answer is that some of the photons, on their way to earth, have been transformed into axions while passing through magnetic fields that populate the Universe. This type of transformation is an allowed process within the structure of any theories in which axions are an ingredient. It is the analogue of the oscillation between neutrino species you encountered in Chapter 4, and it might occur in precisely the amount needed to explain the decrease in supernova luminosity. It is unknown if the transformation actually does take place, much less if axions exist. This conjecture would clearly receive a big boost if dark-matter searches were to detect axions.

The other two explanations of the acceleration are based on conjectures that concern theories of quantum gravity that transcend the Einsteinian paradigm. As I remarked earlier, general relativity does not contain Planck's constant and is therefore a classical physics type of theory, just like the 19th century theory of electromagnetism that underlies the existence of electric motors and generators, household electricity, telephony, etc.

It is widely accepted that any framework that seeks to replace Einsteinian gravity *must* incorporate quantum effects. *String theory* is an example of such a generalization.[16] It is a theory that might be a significant step toward the ultimate ToE. After reviewing a few of its relevant features, I shall describe two string theory–based mechanisms that have been recently proposed as solutions to the acceleration/cosmological constant problem.

Although there are various versions of string theory, the ingredient common to all of them are miniscule vibrating strings, some open, others closed. These theories, rather than being four-dimensional like the spacetime of general relativity, are multidimensional, where "multi" can be 10 or 11 or higher. By extending string theories to include supersymmetry, and assuming that all but four of the dimensions can be "rolled up" in such a way as to be unobservable, the remaining four-dimensional portion has the structure of a quantized theory of gravity, an entity long sought by physicists. Furthermore, when quantum effects can be ignored, the equations reduce to those of general relativity. These facets make string theories highly appealing to their partisans, despite difficulties encountered in attempting to solve the multidimensional equations.

The strings of string theory have some properties in common with ordinary strings, say those on a musical instrument. In the latter case, plucking or bowing the string causes it to vibrate, thereby producing sound; shortening the length of the string leads to notes of higher frequency: recall Chapter 3. These vibrations are *modes of excitation* of the musical string. Correspondingly, the vibrations of the string-theory strings—the only kind of strings I shall consider from now on—are its modes of excitation. Since these vibrations behave like particles, they are taken to be the sources of the standard model particles. A different type of entity is also present in string theories, one that has the same characteristics as a quantum particle known as the *graviton*. Its presence is one of the reasons that superstring theories have caught the imagination of many theorists, for gravitons are the particles that must be present in a quantized theory of gravity.

In quantum field theories, the interaction between a pair of particles is mediated by the emission and absorption of other particles. For example, the emission and absorption of neutral, zero-mass photons by a pair of charged particles gives rise to the electromagnetic interaction between the two charges. The analogue of a photon in the gravitational case is the graviton: it is the neutral, zero-mass particle whose exchanges between two masses produce the gravitational interaction. The presence of gravitons is a crucial test for any theory that could be considered as a potential quantized theory of gravity; string theories pass this test.

These two features of string theory are sufficient background for examining two string theory–based solutions to the acceleration problem. In one, a class of theories denoted *bi-gravity cosmologies*, a non-zero-mass graviton is present. The presence of a so-far undetected massive graviton not only leads to a period of acceleration in the bi-gravity cosmologies, but it has also been suggested as a possible source for inflation in the very early Universe.[17]

A second string-theory mechanism that induces an acceleration of the Universe concerns the extra dimensions. Rather than being rolled up, they are conjectured to behave in much the same way as the three dimensions of our ordinary terrestrial experience. In this non-rolled-up string theory scenario, gravitons can escape from our three-dimensional world into that of the extra dimensions. Once having done so, they leak away forever, thereby altering the usual law of gravity at both short and long distances. The leakage into the extra dimensions leads to an acceleration that behaves as if a cosmological constant were present, but without leading to an initially gigantic Ω_Λ that somehow must be cancelled. The authors of this proposal have identified experiments that could test its validity.[18]

The final conjecture claiming to solve the acceleration problem is an alternative theory of quantum gravity, denoted *loop quantum gravity*. In it, space and time are quantized, thereby leading to a minimum value for areas, volumes, and times. Space and time are thus discrete, rather than being continuous, as assumed is in all other theories. The minimum spatial and temporal values in loop quantum gravity are derived from the Planck length of 10^{-35} meters and the Planck time of 10^{-43} seconds. The theory is constructed from the three spatial and one time dimensions familiar terrestrially: in contrast with string theory, no extra dimensions occur.

Much remains to be worked out in loop quantum gravity, but the following has been achieved: it describes an accelerating universe with a cosmological constant; it reduces to general relativity, but so far only in limited situations; and it reproduces certain black hole results, for example the prediction that black holes can emit radiation. Although the theory is still being developed, there are several tests already proposed that could confirm its quantized spacetime premise.[19] It is possible that loop quantum gravity might eventually become the ToE; however, further development is required before its overall applicability to the Universe can be assessed.

The Singular Nature of General Relativity

It should be clear that despite huge successes, our understanding of the Universe is far from complete: many questions remain unanswered. A major one concerns the Big Bang. Still unknown— and maybe unknowable—are both its true nature and what, if anything, preceded the Big Bang. Unfortunately, general relativity cannot be the path to resolve such questions. The reason is that as the scale factor R approaches zero, that is, at the zero spatial volume wherein the Big Bang supposedly occurred, the equations of general relativity become infinite. Because of this, they provide no information whatsoever. In more technical language, the equations contain a *singularity*. Singularities are the bane of theoretical physics and signal a breakdown of the theory. Their occurrence in a theory means that it either needs to be modified in some way or replaced entirely.

Although the Big Bang concept acknowledges the singular nature of general relativity, one possibility is that its effects might be attributed to the inflationary expansion of the very early Universe [see, for example, Guth (1997)]. But for the purposes of this book, the term *Big Bang* reflects both the singular nature of general relativity—a collapse of the Universe into a volume that becomes tinier and hotter as time runs backwards—as well as ignorance concerning the actual $R = 0$ point.

The existence of the Big Bang singularity is not the only reason that general relativity cannot be the sole description of the Universe, since for microscopic values of R, quantum effects must be taken into account. If the attempts to incorporate quantum effects into a new theory of gravity are ultimately successful, the new theory would almost certainly solve the singularity problem: the discreteness arising from a quantum description should preclude R from becoming zero. Achieving this end is a long-sought goal of elementary particle physicists and cosmologists.

I have already identified two candidate theories of quantum gravity; each contains the discrete element that removes the singularity of general relativity. Discreteness in each is related to the Planck length, the smallest distance that all theoretical frameworks currently recognize.

String theory eliminates the $R = 0$ singularity through its inherent stringiness: instead of elementary particles being point masses, they cannot be smaller than the length of the string whose vibrations create them. Because—as it turns out—quantum theory gives rise to singularities due to the zero size nature of point masses (an $R = 0$ type of occurrence), having all particles larger than a mathematical point eliminates the singularity from string theory. In loop quantum gravity, the quantization of areas and volumes means that none of its particles can occupy a volume smaller than the cube of the Planck length. Non-point mass particles and the absence of singularities are the norm here as well.

Although essential problems are solved in each theory, fundamental questions remain: can either one be the correct quantized theory of gravity? And, if one of them eventually achieves that status, will it also turn out to be the ToE? (Many theorists think string theory either is or will become the ToE.) Furthermore, if this does happen, will it lead to a complete understanding of elementary particle physics and/or of the Universe?

These are questions whose answers lie in the future, if indeed they will be answered at all. In a different vein, the nature of the Big Bang itself is also addressed in each of these frameworks. In the loop quantum gravity scenario, the Big Bang and its subsequent (standard) evolution occurs as a Big Bounce, which follows a previous rapid contraction of the Universe. What might have preceded the contraction awaits solution of the relevant equations, but there is a time earlier than the Big Bang.

In one string-theory scenario, known as the *ekpyrotic model*,[20] the Big Bang occurs as the result of a collision between two structures in the multidimensional framework. These structures, denoted membranes, are usually referred to as *branes*. A collision between two *initially flat*, three-dimensional branes would release gigantic amounts of energy as a fireball in each (like the Big Bang); that energy would not be uniform but would have hotter and cooler spots, like those represented by the CMB anisotropies.

After the collision, three things would occur: first, the two branes would move apart; second, the energy released would lead to structures in the cosmos similar to those of our Universe; third, each would undergo an ever-continuing expansion at an increasing speed. The force causing the accelerating expansion behaves—surprise, surprise—like the cosmological constant. These remarks yield a scenario that reads like one I have outlined in this book. It is plausible, and it may be viable; unfortunately, it has a "slightly" annoying feature: an extra dimension in the ekpyrotic theory experiences its own singularity, and that is a problem that needs to be overcome. In this model as well, there is a time prior to the Big Bang.

Our Universe, Other Universes

There is a further aspect to the ekpyrotic scenario, one that serves as an introduction to the subject of this section. It is the possible existence of other universes, ones additional to ours.[21] In the ekpyrotic model they occur because *two* three-dimensional universes are created as a result of the collision. This is the mechanism that would have created our Universe and another, presumably similar to it. But, because of their motion away from each other, these two universes could not be in contact. Hence, for every pair of flat branes that collide, two universes are created, none of them necessarily identical to ours. A further aspect to this scenario, one shared with certain models of inflation, is the phenomenon of *cyclic* creation. It can happen because the expansion ultimately leads—in the very far distant future—to a structure so large that it is again flat. Then, if there is another flat brane nearby, the two may collide, and the whole process would begin again. Creation of universes in this model might occur many times over.

An alternative mechanism that posits the existence of other universes is an inflationary scenario known as *eternal inflation*, wherein the particular process that causes inflation does not occur uniformly. As a result, an inflationary expansion occurs in some regions before it does in others. The consequence is rapid creation of a whole series of universe-like bubbles that are swept away from one another, eventually leading to the generation of many universes, ours presumably being among them. However, once started, this development will continue indefinitely: small bubbles can inflate nonuniformly, always giving rise to additional expanding and isolated bubbles, each of which become separate universes. There is essentially no end to the creation of universes in this way, and an infinite number would be created. (Eternal inflation is scientifically sound but presumably untestable.)

In each of these situations, there is no reason to think that the universes would have been created in any manner other than randomly. Consequently, in some of them, either the cosmological parameters or the constants in the laws of nature (or both) might not be the same as in our Universe.[d]

One of the implications of having different parameters or constants for a whole class of universes, perhaps an infinite number of them, is that very likely stars, planets, and carbon-based life as we know it would not have evolved in most of them. This may seem like a far-fetched claim, but, in fact, our Universe is finely tuned to the values of these quantities, as many scientists have discussed. Small deviations from them would lead to drastic consequences. Consider, for instance, the energy released when an alpha particle is produced by the burning of four protons in stars.

---

[d]The assumption inherent in this statement, that the laws of nature are universal, is an unimportant one, since I will restrict the discussion to those hypothetical universes where at least the overall mathematical structure of the laws is the same.

As noted for example by Rees (2000), if the ratio of that energy to $4M_\mathrm{p}c^2$ were 0.006 or 0.008 rather than 0.0066 (see page 78), then life as we know could not exist.[22] Changes in other quantities, for example in the strength of the force of gravity or in the value of the cosmological constant, can lead to the same or analogous consequences.

Such fine tuning leads almost naturally to the following question: why, out of the possibly infinite number of universes randomly generated by the above mechanisms (and perhaps others as well), is our own Universe so finely tuned? There are no universally accepted answers to this query, the theological one not excepted. Claiming coincidence is unsatisfying from a scientific as well as a theological standpoint. However, a slightly less unsatisfying answer, though it sheds no light on details, is given by the so-called weak anthropic principle.[e] Loosely paraphrased, it states that because *we* are in it, the Universe can scarcely be different than it is, for slight differences would result in a universe inimical to life. In other words, very little choice of parameters or constants is available if any universe is to be capable of leading to life as we know it.

The weak anthropic principle does not, of course, provide a scientifically satisfying explanation. It borders on the statement "we're here because we're here." It also resembles the Copenhagen interpretation of quantum theory, at least as it is understood by many physicists. Their understanding of this interpretation is that an observer (a conscious human one) must be present for any particular outcome out of the total set of quantally allowed outcomes to occur.[23] Once again, *we* are necessary.

There is also a strong anthropic principle, one some scientists are decidedly uncomfortable with. It may be loosely paraphrased as "any universe must contain those laws of nature that will allow life to develop." In contrast with the blandness of the weak principle, the strong one seems either mystical or theological; neither counts as a scientific explanation.

---

[e] Anthropic principles are touched on in several of the bibliography listings, e.g., Harrison (2000), Livio (2000), Rees (2000), Weinberg (1994) and (2001), plus references cited therein.

In a sense, the Universe may incorporate this latter comment, since at the deepest level it might always be beyond complete scientific understanding.[f] Despite this, it is likely to remain the grandest intellectual game in town, as this book has tried to demonstrate.

---

[f] In this vein, you should bear in mind that despite the optimistic picture presented in this book, scientists, by the nature of their enterprise, are necessarily skeptical folks and so are not congratulating themselves that almost everything fits together well. This is especially true when it comes to assessing the reliability of conclusions based on measurements at cosmological distances. There are uncertainties, assumptions, and systematic errors, some of the latter unknown, that could alter conclusions. A recent review examining these issues is that of R. H. Sanders (2005); however, most of it requires a strong technical background.

Appendix A: Powers of Ten

The power-of-ten notation is a wonderfully compact way of expressing not only very large numbers but also exceedingly small ones. At its core is the fact that numbers are normally expressed using the base 10. As an illustrative reminder, consider the number 234. In it, the digit 4 is in the unit's column, 3 is in the ten's column, and 2 is in the hundred's column. That is, 234 means $2 \times 100 + 3 \times 10 + 4 \times 1$. Each of the three members of the preceding sum is a number times a *power* of 10: $100 = 10 \times 10 = 10^2$, so that 100 equals 10 to the power 2. Similarly, $10 = 10^1$ (i.e., 10 to the power 1), and by definition of the power 0, $10^0 = 1$.

The general case is 10^n, where for the present the *exponent n* (i.e., the *power* to which 10 is raised) is a positive integer or zero. 10^n is a compact way of writing 1 followed by n zeros, for instance $10^4 = 10,000$. Furthermore, *any* number greater than 1 can be expressed this way. A simple example is 40,000: $40,000 = 4 \times 10,000 = 4 \times 10^4$. This notation allows you to express exceedingly large numbers very neatly: it is just a question of determining the correct exponent. Another example is the once-famous gigantic number 10^{100}, denoted a *google*: it equals 1 followed by 100 zeros. The salutary nature of the notation should be evident.

The power-of-ten notation can be extended to include numbers less than one, as follows. First recall that a fraction such as $1/m$ has a specific meaning: it represents one of the portions obtained when 1 is divided into m equal parts (m is assumed to be an integer). When m is a power of ten, the decimal notation comes into play: $1/10 = 0.1$, $1/100 = 0.01$, $1/1000 = 0.001$, etc., where, for reasons of clarity, the decimal point is preceded by a 0. Each of these examples is of the form $1/10^n$, and the general rule for expressing such a number as a decimal is a decimal point followed by $n - 1$ zeros followed by 1. That is, $1/10^n$ means 1 in the nth place to the right of the decimal point. Rather than retain the power of ten in the denominator of this fraction, an altered notation has been introduced: one defines $1/10^n$ as 10^{-n}, which transforms a fraction into a *negative exponent* power of ten. Hence, 10^{-n} becomes a compact, non-fraction means of writing $0.000 \ldots 001$, where the total number of zeros is $n - 1$. If instead of $1/10^n$ one has $m/10^n$, the rule becomes $m/10^n = m \times 10^{-n}$. As an instance of this rule, the very small number five tenths of one billionth, or 0.0000000005, becomes 5×10^{-10}.

You should not regard the introduction of the 10^{-n} notation as a purposeless piece of pedagogy: it serves a highly needful function, since exceedingly small numbers occur again and again in modern cosmology. Two among many examples are the mass of a hydrogen atom, which is about $1.67 \times 10^{-27}\,kg$, and the possibility of the event known as inflation, during which the very early Universe underwent a period of immense expansion lasting about 10^{-34} seconds. (Masses are introduced in Chapter 2; inflation is discussed in Chapter 9.)

The foregoing description is summarized in Table A.1, which also includes the names of the numbers and a few of the common symbols and prefixes in current use. *Mega* and *kilo* are frequently encountered, as in megawatts (MW) for electrical power and kilometers (km) for distance. Ditto for *centi*, whereas *nano*—for example *nanotechnology*, referring to microscopic machines of size $10^{-9}\,m$ or so—has only recently become an element of common usage. An example of a prefix not listed in Table A.1, one

Table A.1. Powers of Ten*

Number	Name	Power	Symbol	Prefix
0.000,000,000,000,001	Quadrillionth	10^{-15}		
0.000,000,000,001	Trillionth	10^{-12}		
0.000,000,001	Billionth	10^{-9}	n	nano
0.000,001	Millionth	10^{-6}		
0.001	Thousandth	10^{-3}		
0.01	Hundredth	10^{-2}	c	centi
0.1	Tenth	10^{-1}		
1	One	10^{0}		
10	Ten	10^{1}		
100	Hundred	10^{2}		
1,000	Thousand	10^{3}	k	kilo
1,000,000	Million	10^{6}		
1,000,000,000	Billion	10^{9}	M	mega
1,000,000,000,000	Trillion	10^{12}		
1,000,000,000,000,000	Quadrillion	10^{15}		

*Adapted from Berman and Evans (1977).

hardly known outside the community of nuclear and elementary-particle physicists and engineers, is *femto*, which means 10^{-15}. It is named after Enrico Fermi, the Nobel Prize–winning physicist for whom the Fermi National Accelerator Laboratory is also named. A *femtometer*, initially denoted a *Fermi*, is 10^{-15} m, a length approximately equal to the radius of a proton, a particle first discussed in Chapters 3 and 4.

Sometimes it is necessary to multiply numbers expressed as powers of ten. The rule for accomplishing this is to multiply the numerical prefactors and add the exponents, as in $10^n \times 10^m = 10^{n+m}$. An instance is the following simple product: $(2 \times 10^2) \times (3 \times 10^4) = (2 \times 3) \times (10^2 \times 10^4) = 6 \times 10^6 = 6{,}000{,}000$. Negative exponents also obey the addition rule: $10^n \times 10^{-m} = 10^{n-m}$. In this case, the power of ten that is the numerically larger number determines whether the result of the multiplication is greater or less than one. Thus, the product of one tenth and one hundred, which equals ten, is $(1/10) \times 100 = 10^{-1} \times 10^2 = 10^{-1+2} = 10^1 = 10$, whereas the product of one hundredth and ten, which equals one tenth, is $(1/100) \times 10 = 10^{-2} \times 10^1 = 10^{-2+1} = 10^{-1} = 1/10$. The addition-of-exponents rule also explains why $10^0 = 1$. To show this, divide any power-of-ten number, for instance 10^n, by itself: $10^n/10^n = 10^n \times 10^{-n} = 10^{n-n} = 10^0 = 1$, as must be, since any number divided by itself is one.

A useful feature of this notation follows from expressing big exponents as a sum of smaller ones, as it often allows a very large number to be stated in ordinary words, for example, in millions or billions. This procedure is activated by searching for ways of writing the exponent as a sum involving the numbers 6 and 9 (or 12, if trillions are desired), since from Table A.1, 10^6 is a million and 10^9 is a billion. For instance, the sun's luminosity, or energy radiated per second (its power), is about 4×10^{26} W, obviously a very BIG number. To get a feeling for how big, let us re-express it in terms of millions and billions by seeking a breakdown of 26 into 6s and 9s. One way to achieve this is via the sum $2 + 6 + 9 + 9$, so that 10^{26} becomes $10^{2+6+9+9} = 10^2 \times 10^6 \times 10^9 \times 10^9$, and therefore 4×10^{26} W is equal to four hundred million billion billion watts!! (That's the same as four hundred trillion trillion watts. Either way, the sun is a bit stronger than any terrestrial light bulb. . . .)

The preceding breakdown into more familiar elements also works for negative exponents. Consider the "size" of a proton, which is about 10^{-15} m (that is, a *femtometer*). This number can be re-expressed in terms of millionths and billionths, as follows: 15 = 6 + 9, from which one gets $10^{-15} = 10^{-6-9} = 10^{-6} \times 10^{-9}$. Hence, a *femtometer* is one millionth of a billionth of a meter. It should be clear that the *femtometer* is a much more appropriate unit for nuclear sizes than the meter.

The preceding comment, on the appropriateness of the meter for nuclear sizes, brings us to the question of how many significant figures are appropriate in stating a number, especially a very large or a very small one when expressed as a power of ten. As noted in Chapter 2, the quantity 10^n generally carries the most important information in a very large or very small number. A case in point is the speed of light, denoted c, whose value is 299,792,438 m/sec. Equivalent statements of this speed are 2.99792438×10^8 m/sec and 2.99792438×10^5 km/sec. There are circumstances where the full nine-digit numbers must be used, but none of them will arise in the context of this book. In general, it is sufficient in the preceding power-of-ten expressions to replace all nine digits simply by 3, so that the speed of light is, to an excellent approximation, either 3×10^5 km/sec or 3×10^8 m/sec. In each of these numbers, the only significant figure turns out to be the numeral 3; the significant *information* resides in the exponent. However, it is unwise to blindly discard all but one of the digits: the factor multiplying 10^n *is* important in some situations. A case in point is the astronomical unit, AU, introduced in Chapter 2. Needful information could be lost if at least its two-digit, approximate, numerical value of 9.3×10^7 miles (or 1.5×10^8 km) were not used.

Appendix B: Primordial Nucleosynthesis

In this appendix, I will examine some of the predictions concerning the relative abundance of the very light nuclei that were produced in the 17-minute period beginning at 3 minutes aBB. By observing nuclei that are believed to be primordial, astronomers and cosmologists found that the empirically deduced abundances are in good agreement with the theoretical predictions, evidence that strongly supports the Big Bang scenario.

I begin with the 1953 prediction of the primordial abundance of the very light nuclei, made by Ralph Alpher, David Follin, and Robert Herman. Starting with a hot, dense mix of neutrons, protons, electrons, photons, and neutrinos all contained in a tiny volume, and using conventional theories of nuclear reactions and beta decay, they concluded that no neutrons would remain after the primordial nuclei were produced. These nuclei consist of deuterons (d), the helium isotopes ^{3}He and ^{4}He, and the lithium isotopes ^{6}Li and ^{7}Li.[a] They also concluded that as a result of the electrical repulsion acting between the preceding five nuclei, and the absence of stable nuclei with total numbers of neutrons plus protons equal to five and to eight, none heavier than ^{7}Li would have been formed. (The same problem of producing heavier nuclei in stars, and Fred Hoyle's solution to it, is described in Chapter 4.) Finally, they made a prediction of the relative abundance.

Although the above scenario has subsequently been amplified and somewhat modified, and the times marking the beginning and end of primordial nucleosynthesis have been specified with increasing accuracy over time, Alpher, Follin, and Herman's fundamental conclusions have not changed. There are many more details in the analysis than is appropriate to include here, so that my treatment is relatively "broad brush." For additional aspects,

---

[a]See Chapter 4, page 76, for the definition of isotope and Figure B.1 for the neutron/proton structure of ^{3}He, ^{4}He, ^{6}Li, and ^{7}Li.

readers are recommended to the classic monograph by Weinberg (1977).

The starting point of the discussion is 1 second aBB, when positron annihilation ends. At this time photon (and antineutrino) energies have become too low to produce particle–antiparticle pairs, and therefore the numbers of neutrons, protons, and electrons are fixed. Theoretical analysis indicates that there were then about five protons for every neutron. Since these two particles were occupying the same volume, cosmologists express this latter result in terms of number densities (i.e., the number per volume). Number density is designated by the subscripted lowercase letter n, for example n_n and n_p, where the subscripts denote, respectively, neutron and proton. In terms of the ns, one neutron per five protons becomes the ratio $n_n/n_p = 1/5 = 0.2$.

It took approximately another 3 minutes for nucleosynthesis to begin, when T_{ph} had decreased to 7×10^8 K.[b] During this approximate 3-minute interval, the only process of relevance to my story is neutron beta decay,

$$n \rightarrow p + e^- + \bar{\nu},$$

first encountered in Chapter 4 (see page 73 for definitions of the symbols). Three minutes is long enough for some of the neutrons to undergo beta decay, leading to an increase in the approximate number of protons per neutron from about 5 to 7. The number density ratio thus goes from roughly 1/5 to around 1/7; that is, at 3 minutes aBB, $n_n/n_p \cong 1/7 = 0.14$.

Although beta decay was the cause of the early Universe beginning to lose its free neutrons at 1 second aBB, 3 minutes later nuclear reactions took over that role. In other words, T_{ph} had fallen sufficiently low that stable nuclei could begin to form, the first of which were deuterons. They were produced by the capture of neutrons by protons, leading to a final state of photons and deuterons. In terms of symbols, the neutron capture reaction is written as

---

[b]Times and temperatures are from the analyses of Weinberg (1977), Peacock (1999), and Rich (2001).

$$n + p \rightarrow d + \gamma,$$

where d denotes the deuteron [d = (n + p), the neutron–proton bound state] and γ, the lowercase Greek letter gamma, represents the photon; γ, introduced here for the first time, is the standard symbol for a photon.

Once deuterons were formed, other reactions could occur, the end results of which were formation of the series of very light (and stable) nuclei first identified by Alpher, Follin, and Herman. These nuclei and their neutron/proton constituents are listed in Figure B.1, along with the reactions that produce them. A

During the interval between approximately 3 and 20 minutes aBB, a prediction of Big Bang cosmology is that neutrons and protons in the early Universe will initiate a series of nuclear reactions whose stable nuclear products are the very light nuclei. The major reactions that produce them are listed below.

Major reactions:

$$n + p \rightarrow d + \gamma$$

$$d + d \rightarrow {}^{3}He + n$$
$$d + d \rightarrow t + p$$

$$d + {}^{3}He \rightarrow \alpha + p$$
$$d + t \rightarrow \alpha + n$$
$$d + d \rightarrow \alpha + \gamma$$

$$d + \alpha \rightarrow {}^{6}Li + \gamma$$
$$t + \alpha \rightarrow {}^{7}Li + \gamma$$

Notation:

γ = a photon
n = a neutron
p = a proton
d = a deuteron, the nucleus formed from a neutron and a proton: d = (n + p)
 t = tritium, the nucleus comprising two neutrons and a proton: t = (2n + p)
${}^{3}He$ = helium 3, the nucleus formed from a neutron and two protons: ${}^{3}He = (n + 2p)$
 α = ${}^{4}He$ = helium 4, the nucleus with two neutrons and two protons: $\alpha = (2n + 2p)$
${}^{6}Li$ = a lithium isotope, containing three neutrons and three protons: ${}^{6}Li = (3n + 3p)$
${}^{7}Li$ = a lithium isotope, comprising four neutrons and three protons: ${}^{7}Li = (4n + 3p)$

Figure B.1. Primordial nucleosynthesis.

consequence of producing these nuclei is a complete absence of free neutrons in the early Universe: all have either beta-decayed or been used up in forming the nuclei listed in the figure.

The grouping in the figure is *via* the mass of the heaviest nucleus in the final state, which is seen to be the target nucleus in the next group. The reaction in which a proton captures a neutron is necessarily the first in the list, because the next pair, which underlie almost all the rest, cannot proceed without the presence of deuterons. Although most of the ^{3}He produced in the d + d reaction goes into forming αs, a tiny percentage of them survive, whereas none of the tritium will: those not used in making αs or ^{7}Li will beta decay into ^{3}He plus an electron and an antineutrino.

The third group consists of the three α-particle forming processes, all initiated by collisions with deuterons. Alphas are the easiest to make, essentially because of their very large binding energies: the differences between the masses of the particles that form the final, heavy nucleus and that of the final particle is greatest for the α particle. One result of the ease with which αs are formed is that, apart from protons, they are the most abundant of the primordial nuclei.

The final pair shown in Figure B.1 comprises the reactions that produce the lithium isotopes. In analogy to what is found on the earth and in stars, primordial ^{7}Li is produced in greater abundance than is ^{6}Li, though the presence of each is almost nil. Furthermore, primordial ^{6}Li, like primordial ^{3}He, is very hard to detect, and so neither of their abundances can be compared with measured ones.

Figure B.1 suggests that all the deuterons will be used up in forming the heavier nuclei, and theory bears this out: there are roughly 10^5 times as many protons as deuterons; measured by mass, protons also outnumber αs, by about four to one.

Nucleosynthesis could begin only when T_{ph} had dropped sufficiently low that photons would not immediately dissociate any of the deuterons into their elementary constituents. Why it didn't continue beyond approximately 20 minutes aBB, when T_{ph} was approximately equal to 3.5×10^8 K, is only partly a result of the lack of stable nuclei whose total numbers of neutrons and protons equaled 5 or 8 (discussed on page 82).[c] Another factor is the decrease in T_{ph}, which triggers a corresponding decrease in the speeds of the nuclei. As these speeds drop, there is a sharp drop in the probability that any pair of nuclei can become intimate enough to permit fusion to occur. The end result is that quantum tunneling through the electrical repulsion barrier becomes negligibly small.[d] Hence, the formation of nuclei ceases.

The preceding discussion is a qualitative description of primordial nucleosynthesis. It is an offshoot of a quantitative analysis, one result of which is the aforementioned prediction of relative abundances, that is, of the ratios formed from each of the numbers of d, ^{3}He, ^{4}He, ^{6}Li, and ^{7}Li to the number of protons, all at 20 minutes aBB.

The numerical values of these ratios are dependent on the conditions in the Universe at the approximate age of 3 minutes aBB, in particular on the relative number of neutrons plus protons to photons. The number density of neutrons plus protons is conventionally denoted n_B (B standing for baryon), while that of photons is n_γ. Rather than write out the ratio n_B/n_γ each time they use it, cosmologists prefer to express it by the symbol η, the lowercase Greek letter eta: $\eta = n_B/n_\gamma$. However, the value of η is unknown, and it therefore becomes a parameter in the equations that determine the theoretical number densities n_d, n_3, n_α, n_6, and n_7, where the subscripts denote, respectively, deuterons, ^{3}He, alpha particles, and the nuclei ^{6}Li and ^{7}Li.

The number densities, and their resulting abundances, have been determined for a large range of η values, typically 10^{-12} to 10^{-6} (there are obviously many more photons than protons). A plot of the calculated abundances shows that deuteron abundance is the most sensitive to the value of η: the other four abundances display far less variation as η is varied. For example, from

---

[c]If, for example, ^{5}He = (3n + 2p) and ^{8}Be = (4n + 4p) *were* stable, they could be stepping-stones to the formation of some of the other not-so-very-light nuclei.

[d]Tunneling is discussed on page 76 in the context of the reaction p + p $\rightarrow$ d + e$^+$ + v.

analyses predating that of the WMAP data, η is expected to lie between 10^{-10} and 10^{-9}. As η increases through this latter range, n_α/n_p is almost constant; the maximum to minimum values of n_7/n_p is about a factor of 10; and n_d/n_p falls by 100. Furthermore, as η changes from 10^{-12} to 10^{-8}, n_d/n_p decreases by a factor of roughly 10^9, far greater than the changes in any of the other relative abundances.[e]

In view of these numbers, it should be no surprise that the value of η has been fixed by using the measured value of n_d/n_p. (Experiments attempting to measure the ratio n_d/n_p have been an important activity for some time.) Once a value of η has been determined, the theoretical abundance of the rest of the nuclei can be read off the abundance *versus* η plot and then compared with measured values from what are believed to be either primordial sources or extrapolations to them.

The ratio of the first to the second acoustic peak values in the power spectrum (Chapter 7) determines n_d/n_p, which is the abundance at approximately 379,000 years aBB. On the other hand, measurements of this ratio from quasars, which are believed to be the best source of primordial deuterium, yield values from essentially 3 minutes aBB. These power spectrum and quasar sources should yield the same answers, since there should be no change in the ratio from the time of primordial production to 379,000 years aBB. The combined WMAP result is $n_d/n_p = (2.62 + 0.18 - 0.20) \times 10^{-5}$, whereas the primordial measured value—as of late 2003—is $(2.78 \pm 0.29) \times 10^{-5}$, which encompasses the preceding theoretical value. The agreement between the values obtained from such different times is powerful evidence supporting Big Bang cosmology. The combined WMAP value of n_d/n_p leads, omitting all uncertainties, to $\eta = 6.1 \times 10^{-10}$, a result consistent with other values (e.g., 5.9×10^{-10}). From a plot of abundance *versus* η, one then finds $n_7/n_p \cong 4 \times 10^{-10}$. In place of n_α/n_p, however, theorists prefer to calculate the primordial mass fraction of αs to that of αs plus protons, the symbol for which is Y_p. The same plot yields a calculated value of $Y_p \cong 0.248$.

---

[e]Readers interested in the abundance *versus* η plot should be able to find it on the WMAP Web site.

Measurements of Y_p yield values that cluster around 0.24 (uncertainties are absorbed into this number), and those of n_7/n_p lead to the range estimate $(1.23 - 2.19) \times 10^{-10}$ (with large possible corrections from certain uncertainties intrinsic to experimental extrapolations). The extremely small amounts of ^{3}He and ^{6}Li created primordially make them too difficult to measure, as noted above, and so are not listed. As advertised, the agreement between the calculated and measured numbers, though not perfect, is certainly good enough to justify the claim that Big Bang nucleosynthesis, and thus Big Bang cosmology, is well supported experimentally.[f]

---

[f]Agreement to within a factor of 4 or so between the theoretical and experimental values of n_7/n_p could be considered a triumph, given the extreme sensitivity of the theoretical values to η and the difficulty of measuring primordial abundance ratios. Future experiments should improve the measured values and thus provide a somewhat more sensitive test. But there seems to be little question that the majority of cosmologists believe that the results are strongly supportive of Big Bang cosmology. To have been able to make, as well as verify, claims about the condition in the Universe at such early times in its history surely ranks among the most profound of all human achievements. You may wish to contemplate other possibilities in this regard.

Appendix C: The Elementary Particle Zoo

The first steps on the path leading to the determination of the elementary constituents of matter were taken in the last five years of the 19th century. A total of three unknown kinds of radiation, named alpha, beta, and gamma rays, had been identified in that interval. Beta rays were shown in 1897 to be the same as those negatively charged particles that their discoverer, J.J. Thomson, had denoted *electrons*. Beginning with this result, it took about 15 years more for the other two unknowns to be identified as helium nuclei and electromagnetic radiation.[a]

By the time of the Second World War, atoms were known to consist of electrons and a much more massive nuclear core, nuclei had been shown to contain neutrons and protons, positrons had been postulated to exist and were then discovered, neutrinos had been conjectured to exist, and a strongly interacting particle called a *meson* had been detected.

After the Second World War, the use of particle accelerators rapidly accelerated the process of particle discovery, and by the 1960s, hundreds had been detected. They were divided into two classes: *leptons* and *hadrons*. The former contains electrons and neutrinos plus their more massive relatives (about which more below) and the antiparticles to these objects. Hadrons, particles that interact strongly with one another,[b] were further subdivided into the two categories of mesons and baryons, the latter of which was introduced in Chapter 8.

In a very real sense, the discovery of hundreds of strongly interacting particles was an embarrassing and frustrating richness:

---

[a]For some details of their discovery and identification, see Weinberg (1983).

[b]"Strong," in contrast with *electromagnetically* interacting electrons, and *weakly* interacting neutrinos.

were there really so many "fundamental" objects? This question led to efforts to describe the large variety of hadrons in terms of more elementary constituents, in much the same way that nuclei are described by enumerating their different numbers of neutrons and protons.

Success has been achieved via *quantum chromodynamics,* the framework whose fundamental particles are *quarks* (and *gluons,* described below). Quarks are particles with highly unusual properties, such as fractional charges. Quark substructure was a scheme initially postulated for taxonomic purposes only, but experiments performed at very high energies established that protons had a tripartite substructure, which was eventually shown to be consistent with quarks. Much theoretical analysis, including Nobel Prize–winning work, gave rise first to the framework known as the standard model of elementary particle physics and then to a number of further developments, some of which are briefly discussed in Chapter 9.[c]

The standard model melds the electroweak theory and quantum chromodynamics into a theoretical structure comprising 17 elementary particles and the equations that govern their interactions. The particles are divided into three classes, consisting of six quarks, six leptons, four *bosons,* and a fifth, hypothetical boson known as the *Higgs,* named for the physicist who proposed it. These particles are shown in Table C.1.

Table C.1. The 17 elementary particles of the standard model

Quarks		
u: up	c: charm	t: top
d: down	s: strange	b: bottom
Leptons		
e: electron	μ: muon	τ: tau
ν_e: e neutrino	ν_μ: μ neutrino	ν_τ: τ neutrino
Bosons		
γ: photon Z: Z boson	W: W boson	g: gluon
Higgs		

---

[c]Readers who wish to learn about the history of these topics might consult Riordan (1987).

The six members each of the quark and the lepton categories are identified both by their standard symbols and names. The groupings in the table reflect the current understanding, in which the quark and the lepton sets are divided into three separate families. Each family pair is represented by one of the two-element columns.

At one time, it was thought that the only quarks were the *up*, the *down*, the *charm*, and the *strange* (names bearing historical significance), but later theoretical analysis suggested a third family. Its members, given the names *top* and *bottom*, were detected by massive experimental efforts. Similarly, the electron and its heavier sibling the *muon*, along with their neutrino partners, were initially thought to be the four members of the electron family, but discovery of the *tau* brought the membership from four to six. A major difference between quarks and leptons is that all the quarks are believed to be stable (at least to within a lifetime of about 10^{33} years), whereas both the muon and the tau are unstable. The mean lifetime of their various decay modes are about 2×10^{-6} seconds for the muon and roughly 3×10^{-13} seconds for the tau.

Quantum fields are entities mentioned in the main text: see the footnote on page 203 and especially the comments on page 217, where the emission and absorption of particles that mediate interactions is characterized. In the standard model, the exchanged particles that mediate the interactions are denoted *bosons*; they comprise the fifth row in Table C.1. Photons are associated with the electromagnetic field, while the Z and W bosons mediate the weak interaction. Gluons are the "glue" that binds quarks together in hadrons, the strongly interacting particles. However, in quantum chromodynamics not only do quarks interact by means of gluon exchange, but the gluons also interact with each other. This latter feature is a major reason why this framework is so complex and its equations so difficult to solve, for instance when it is used to describe the deuteron as an interacting system of six quarks.

To help understand why a deuteron, the nucleus consisting of a neutron and a proton, is also a six-quark system, I consider next the quark structure of neutrons and protons (and later the quark structure of one of the many mesons). The six-quark structure follows from the fact that neutrons and protons are each three-quark systems composed of up and down quarks (u and d, respectively). In order that the various quantum aspects of these two nucleons be correctly accounted for, and in addition that the neutron has zero charge while the proton has one unit of positive charge, the u and the d, like their other four siblings, must have highly unusual properties, as I indicated above.

The unusual quark property I shall concentrate on is their possession of *fractional* charges (no other objects are known to share this feature). In units of the elementary charge, that of the up quark is 2/3, whereas that of the down quark is −1/3; the *signs* but not the magnitudes of these numbers are changed in the antiquarks $\bar{u}$ and $\bar{d}$. When forming a particle from a group of quarks, the charge of the composite particle is the *sum* of the charges on its constituents. Protons therefore consist of two u quarks and one d, as they are the only tripartite combination whose charges add up to one unit. Correspondingly, neutrons are composed of two d and one u, since the sum of quark charges here adds to zero. In terms of symbols, p = (u + u + d), while n = (d + d + u), where the parentheses indicate a particle-stable, bound configuration. While it might be thought, in analogy to the meson case discussed below, that $\bar{u}$ and $\bar{d}$ should also enter the quark substructure of the nucleons, antiquarks are ruled out because of other properties that I have not addressed.

The presence of three quarks in each nucleon means that various quantum attributes of the quarks can be combined in different ways, leading to *excited states* of neutrons and protons. Such nucleon excited states, the precise analogues of the excited states of atoms described in Chapter 3, have been detected experimentally—they are among the hundreds of what were once thought to be the "elementary particles" remarked on above.

Mesons, in contrast with neutrons, protons, and their excited states, are particles formed from *two* quarks. An example is the *pi* meson, denoted by the lowercase Greek letter π. Often called *pions*, they can have positive, negative, or zero charge. Their symbols are, respectively, π^+, π^-, and π^0. Pions are composed of up and down quarks, as well as their antiparticles $\bar{u}$ and $\bar{d}$. Using the same symbology as for nucleons, the quark substructure of pions is $\pi^+ = (u + \bar{d})$, $\pi^- = (d + \bar{u})$ and $\pi^0 = [(u + \bar{u}) + (d + \bar{d})]$. The neutral pion has a more complicated structure than the other two because zero charge can be formed in the two ways specified by the quark–antiquark pairs in the two sets of parentheses.

These examples illustrate some of the ways quarks and antiquarks can be combined to form hadrons, and I turn next to the remaining particle in Table C.1, the Higgs boson. Although still hypothetical, the Higgs has long been endowed with the mantle of reality, since in the standard model, it is the particle from which all the other ones obtain their mass. That is, without the quantum field whose interaction with the other 16 particles is mediated by the Higgs, those 16 would have zero mass. Since each of them does have mass, the Higgs, as the mass giver, has been referred to as the "God" particle. Its detection, like that of a dark-matter candidate, has motivated (difficult) experimental searches, although no definite results have been obtained.

A tantalizingly indefinite result has been reported, however. It is from Higgs-search experiments conducted at the now-dismantled Large Electron Positron (LEP) accelerator once operated at CERN in Switzerland. The tentative finding is of a particle whose mass is about 115 times that of the proton mass, a value consistent with various expectations for the Higgs mass. Further experiments looking for the Higgs boson await the upgrade of the Tevatron accelerator at the Fermi National Accelerator Lab in Illinois and the completion of the Large Hadron Collider (LHC) at CERN. The ultimate detection of the Higgs is crucial to the continued acceptance of the standard model; it would certainly confer a degree of glory to the team that discovers it. Interested readers should stay alert for news, starting as soon, perhaps, as 2007 or 2008.

Chapter Notes

Chapter 2: Measuring Distances

1. The formula, from plane trigonometry, is $h = D \times \tan A$.
2. Deviations from sphericity—the equatorial bulge and the presence of other small deformations—are discussed in books on astronomy, for example, Abell (1975).
3. There are many treatments of these topics. Two that the author has enjoyed are the books by Koestler (1959) and Kolb (1999).
4. See, for example, the recent book by Teresi (2002).
5. More details on the determination of the earth's radius and on the earth–moon and earth–sun distances can be found in Webb (1999).
6. For discussions of Copernicus, Kepler, and Newton, readers might consider the books mentioned in note 3, as well as Kuhn (1957). More technical comments can be found in Abell (1975) and Webb (1999).
7. Because of the presence of other planets, all their orbits are perturbed slightly away from elliptical. The deviations are small and calculable; it was the slight deviation of the orbits of Uranus and Neptune that led, respectively, to the discoveries of Neptune (by Johann Galle in 1846) and Pluto (by Clyde Tombaugh in 1930). Recently, giant ice-balls have been discovered beyond the orbit of Pluto; if they are not accorded planetary status, Pluto could conceivably be downgraded from a planet to an iceball.
8. As defined by Figure 5, the determination of D is *via* the procedure is known as horizontal parallax, and A is the parallax angle. More generally, the angle of parallax is defined as the angle between the lines drawn from the two observation points that meet at the object.
9. A "fixed" star is so far away that its motion can be ignored, even though its movement might be inferred from the earth or with orbiting telescopes.
10. The name for g is the "acceleration of gravity." The value of g on the earth is about 9.8 meters per second per second or roughly 32 feet per second per second. Acceleration is the rate of change of speed with time; hence the presence of two "per seconds." For the moon, these numbers decrease by about one sixth, while for the sun they increase by roughly a factor of 28.

11. The value for $d_{\text{Lum Matt}}$ in Table 7 is derived from the "Big Bang Nucle-osynthesis" section of the 2004 Review of Particle Physics issue of *Physics Letters* B, listed in the Bibliography. The relevant quantity listed in the review involves the Hubble constant H_0, for which I have used the value of 71 km/sec per Mpc (see Chapters 6 and 7 for defi-nitions and discussion of the Hubble constant).

Chapter 3: Light, Radiation, and Quanta

1. See Park (1997) for a discussion of this and related points.
2. As presented in the text, relation (1) is the nonrelativistic approxi-mation to Doppler's formula for the case of electromagnetic radia-tion. The exact formula is

$$[(\lambda_{\text{obs}} - \lambda_{\text{emit}})/\lambda_{\text{emit}}] = \{[(1 + V/c)/(1 - V/c)]^{1/2} - 1\},$$

where the one-half power in the symbol $[\ldots]^{1/2}$ means square root.
3. A delightfully droll account of the Buijs-Ballot experiment is given by Kolb (1999).
4. Park (1997) briefly discusses some of this work and has references to it; see also Weinberg (1983). A technically demanding source is Whittaker (1951).
5. See, for example, Park (1997) or Gregory (1976).
6. Some heated solids give off a continuous spectrum of radiation; gases typically radiate bright-line spectra. See Kolb (1999) for a discussion of these matters, and especially for an amusing but disheartening anecdote concerning the impact that Gustav Kirchoff's work failed to make on at least one learned nonscientist.
7. Park (1997) points out that it takes only 5 to 7 photons to excite the receptors in the human eye.
8. Feynman (1985) treats not only quantum aspects of light but also of matter. His book is somewhat demanding but quite rewarding. See also Park (1997) and Rigden (2002).
9. The words *numerically least* are used because the internal energies of systems that contain two or more bodies bound together in a stable configuration are all *negative*. In particular, the lowest energy of a microscopic system takes on the largest negative value, with the higher internal energies all having numerically smaller but still neg-ative values. For discussions of these topics see Born (1957)—though this book is for the technically-proficient reader—as well as Feynman (1985), Park (1997), or Weinberg (1983).

Chapter 4: Stars

1. Red giants are a stage in stellar evolution and are discussed in the section on the Hertzsprung–Russell diagram. More details than are presented there may be found, for example, in Webb (1999) or Abell (1975).
2. The relation between °F and °C is given by the formula °F = 32 + (9/5) × °C.
3. The average speed of oxygen molecules at room temperature is approximately 25 m/sec.
4. The data on which Figure 11 is based may be found at the Goddard Space Flight Center Web site, as presented by Dr Robert F. Cahahan: http://climate.gsfc.nasa.gov/~cahahan/Radiation/SolarirrVblackbody.html. Note, however, that the data is plotted in terms of wave number, not wavelength. The relation is wave number = 2π/wavelength, where $\pi \cong 3.14159$ is the ratio of a circle's circumference to its diameter.
5. A somewhat detailed identification of the absorption dips may be seen at http://climate.gsfc.nasa.gov/~cahahan/Radiation/SolarIrr.html; the data is plotted in terms of wave number (see note 4).
6. The sun's luminosity is $L_{Sun} = 3.85 \times 10^{26}$ W. If all the energy was in the form of photons of wavelength 500 nm (i.e., the wavelength at which the maximum in the sun's radiation curve occurs), approximately 9×10^{32} such photons would be radiated per second.
7. A schematic representation of the absorption and scattering of the sun's radiation due to the earth's atmosphere can be found at the Web site http://rredc.nrel.gov/solar/pubs/shining/page7a_fig.html.
8. See Phillips (1994), Chapter 1, especially pp. 5–7 and 19–20.
9. As long as the radioactive sample is sufficiently massive, it can remain a source of danger after at least several lifetimes. For example, one quarter of the original amount is present after two lifetimes, one eight after three lifetimes, etc.
10. For an excellent account of the early discoveries/work on subatomic particles such as beta particles, etc, see Weinberg (1983).
11. Tunneling is a quantum process that "common sense" deems impossible, until one recalls that common sense is based on events in the ordinary, macroscopic world of everyday events. The underlying idea is straightforward to explain by analogy, although it still seems to defy common sense. Imagine two neutrons speeding directly toward each other on a straight-line path. Since they are electrically neutral, there is no Coulomb repulsion between them, and in

principle they can collide. Next, replace the two neutrons by a pair of protons also speeding toward each other on a straight-line path. Because they are each positively charged, there is a repulsive force between them that increases in strength as they get closer. For the purpose of this discussion, the repulsion—or repulsive barrier— between them will be represented by a high, narrow hill that lies directly in their path. I assume that the speeds of each are insufficient to allow either to roll up to the top of the hill and then roll down. Therefore, it would seem that they cannot collide; the barrier hill prevents it. But, since they are microscopic particles whose behavior is governed by the laws of quantum theory, it turns out that there is a nonzero probability that permits them to pass *through* the hill! The effect of this quantum probability (for barrier penetration) is to cause the barrier to behave *as if* there were a tunnel built through it, along which the two protons could continue their straight-line path and ultimately collide. Hence the phrase *quantum tunneling*.

12. See Appendix A of Webb (1999) for an explanation of astronomical nomenclature.

13. The value of $3M_{Sun}$ for the approximate upper limit to a neutron star's mass has been challenged in papers by H. A. Bethe and G. E. Brown, whose calculations yield $1.5M_{Sun}$ as the limit. Measurements of neutron star masses from binary situations have found the value $1.4M_{Sun}$, consistent with the Bethe–Brown upper limit. However, there is a 2001 paper stating $1.86M_{Sun}$ as the measured value of a neutron star's mass, although an author of the 2001 paper has told Brown that a mass of $1.4M_{Sun}$ is not entirely excluded. You can consult Brown (2005) for a few more details and references to the literature, and you may wish to keep an eye open for announcements of future developments. N.B.: The end stage core and stellar masses listed in the text are amalgams from Webb (1999), Harrison (2000), and Silk (2001).

14. Eddington's inability to accept Chandrasekhar's work is not an isolated instance of reputable (i.e., famous and brilliant) scientists denigrating the work of others—sometimes their own (!)—that they had misunderstood. Eddington did it more than once, as did Albert Einstein. Einstein, who invented the concept sometimes referred to as *dark energy*, which gives rise to the acceleration of the Universe, later referred to it as his "greatest blunder." See Chapters 5 and 7. A detailed study of the Eddington–Chandrasekhar controversy has recently been published by A. I. Miller (2005). It traces the

later decline of Eddington's reputation and the enhancement of Chandrasekhar's, who, according to Miller, never transcended the feelings caused by Eddington's derogatory remarks.

15. The effects of "gravity" studied by Schwarzschild are those due to general relativity, wherein the Newtonian concept of gravity as an unexplained force is replaced by the warping of space and time by a body's mass. This distortion (warping) of spacetime causes a second body moving in its vicinity to alter its path. For objects of relatively low mass, the alteration of the path is just what Newton's theory predicts. See Chapter 5 for further discussion.

16. The radius of the event horizon is given by the formula $D_{Sch} = 2GM/c^2$, where M is the value of the point mass, c is the speed of light, and G is Newton's gravitational constant (it is common to both Newtonian and Einsteinian gravity; readers interested in its value are referred to the books by Abell or Harrison, or to an elementary physics text).

17. Whether a black hole can shrink to nothing is an open question. Two reasons it may not are the possibility of its being charged (where would the charge go?) and the need for a quantum description when its size becomes roughly 10^{-35} m, the so-called Planck length (see Chapter 9 for discussion of this length). The history of the infinite shrinkability of collapsing stars is recounted by Miller (2005).

18. In both the special and the general theories of relativity, space and time are inextricably combined. One result is that observers moving relative to one another will measure both time duration and spatial distances differently.

19. The possibility that black holes would never entirely evaporate is addressed in note 16 above. Furthermore, rotating (spinning) black holes can also radiate. For discussions of this latter feature see Thorne (1994), Rees (2000), or Harrison (2000).

Chapter 5: The Expanding Universe

1. See, for example, Harrison for a discussion of the motions and the final number of 600 km/sec.
2. See Kolb and Webb for discussion of the Curtis–Shapley debate and for biographical material.
3. Webb and Kolb present interesting and useful information on the discovery of spiral nebulae and their distances.

4. The "back-to-back pair of fried eggs" analogy is from Webb.
5. The information on the size and shape of our Galaxy is from Webb. A sense of the huge extent of the Galaxy may be obtained by modeling the earth by an aspirin tablet (0.63 cm, p. 23), relative to which the diameter of the Galaxy is about 897,000 km!
6. In 1928, prior to Hubble's writing out relation (1), Howard P. Robertson, an American general relativist, showed that previously obtained redshifts plus some of Hubble's galactic distance measurements supported a linear redshift–distance relation, equivalent to a speed–distance relation for small redshifts. (The redshift parameter is defined on p. 44). Robertson's "rough verification" was published in the *Philosophical Magazine*, a journal that few astronomers read, and was thus unknown to Hubble.
7. It turns out that $1/h_0$ and $1/H_0$ are estimates of the age of *our* Universe, as discussed on page 109 (and are also the exact age of a universe dominated by radiation).
8. Biographies of Einstein deal with the 1919 deflection-of-light measurements and his accession to international celebrity status. Webb has a nice summary; readers who do not mind the presence of mathematics and technical discussion are highly recommended to the scientific biography of Einstein by Pais.

Chapter 6: Homogeneous, Isotropic Universes

1. General relativity has reached paradigmatic status as a result of its predictions being substantiated experimentally. In addition to those already described and some cosmological ones noted later in the chapter, there are other noncosmological ones. Two of these concern its application to the following situations: measuring the difference in time as recorded by clocks carried in two airplanes that circumnavigated the globe in opposite directions, and the almost continual updating of clock synchronization in the global positioning system (GPS) due to space-distortion effects. The latter situation is beautifully described by Ashby (2002). He notes that the time accuracy of better than 1 part in 10^{13} provided by atomic clocks at the U.S. Naval Observatory is absolutely essential. With it, GPS positions are reliable to within 8 m, but without such incredibly precise timekeeping, which is the cornerstone on which corrections due to general relativity are based, GPS positions would be in error *each day* by more than 11 km!

2. Although rare, such overlooked predictions occasionally dot the scientific landscape. Another example is the phenomenon known as the Ahararonov–Bohm effect, after the two physicists whose 1957 publication alerted the physics community to the theoretical possibility of a new and totally unexpected type of interference phenomenon, analogous to that of Chapter 3. Overlooked by the community was the discussion of this phenomenon 10 years earlier by W. Ehrenberg and R. W. Siday. (The latter authors were acknowledged by Aharonov and Bohm.) Yet a further instance of this fading into limbo was the original work on genetics by the Austrian monk Gregor Mendel. He published his results in an obscure journal in 1866, where they remained unread by biologists until several of them independently discovered it in 1900. Mendel, like Ehrenburg and Siday, received the acknowledgment his work deserved, though long after his death.

3. It is this type of situation that underlies the old joke wherein one of the fishermen aboard the boat announces to the other that he has put a mark on the boat's side to identify the place where the fishing was so good!

4. Readers aware of the dictums of special relativity may object to this simple arithmetic addition of velocities (speeds). There is no problem with such addition, however, since it is being done in the context of an accelerating environment, to which special relativity does not apply. The latter theory describes effects on motion due to observations made from reference frames moving at constant speed relative to one another. General relativity actually allows speeds to exceed the speed of light, which is the limit encountered in special relativity. See, for example, comments in Harrison's book.

5. A possible objection to this conclusion is that the derivation of the speed–distance law is apparently dependent on the special linear configuration of Figure 21. Isotropy, however, guarantees that the conclusion is independent of the orientation of the configuration. Keeping the initial distances the same—whatever it is chosen to be— requires the speeds at which adjacent galaxies are separating to be the same, thereby allowing the observer on A to deduce that C's speed is twice that of B.

6. For readers who have studied calculus, the mathematical relationship is $H(t) = (dR/dt)/R$, where t stands for time and d/dt is the first derivative with respect to time.

7. The two equations obeyed by R are usually referred to as the Friedmann–Lemaitre equations. One is for the quantity $(dR/dt)^2$, the other for the second derivative, d^2R/dt^2. Readers interested in reading up on this are recommended to the books by Harrison, Webb, Berry,

Raine and Thomas, or Rich. However, none is easy going with regard to this topic as well as the many others that are dealt with. Note also that some of these authors use $a(t)$ in place of $R(t)$ for the scale factor.

Chapter 7: The Parameters of the Universe

1. See note 7 of the preceding chapter.
2. The following comment is for mathematically prepared readers: the reason that a nonzero (and positive) value of Λ causes the Universe to expand indefinitely is that Λ will eventually dominate over matter, at which time the scale factor becomes proportional to the exponential of Λ: $R \propto \exp[(\Lambda/3)^{1/2}t]$, where $(\ )^{1/2}$ means "square root" and t is the time. Since Λ and t are both positive quantities, the exponential function grows without bound as t increases indefinitely.
3. The COBE results are described in two of the Web sites listed in the Bibliography.
4. The far-infrared and "near" microwave portions of the electromagnetic spectrum overlap. See Table 8 or Figure 8.
5. Readers interested in how well the data fit the $T = 2.726$ blackbody curve should go to the first of the two COBE Web sites and then click on the "CMB intensity plot."
6. For details beyond those recounted here, see the COBE and WMAP Web sites listed in the Bibliography.
7. There are technical assumptions that are used to generate the Legendre polynomial expansion. See note 8.
8. Readers with a good mathematics and physics background who wish to learn more about the power spectrum and/or the extraction of parameter values might try the Universities of California and of Chicago Web sites (G. Smith and W. Hu, respectively) or the books by Raine and Thomas, Rich, or Pcacock listed in the Bibliography, as well as the references cited therein.
9. The parameter ℓ, against which the peaks and valleys in the power spectrum are plotted, is related to the variable angle A at which the detectors measure temperature differences. The relation is $A \cong 180°/\ell$, where A is in degrees. The peaks occur at $\ell \cong 250, 550, 850$, up to about 1400, and possibly beyond. See Plate 3. Note, by the way, that $\ell \cong 250$ corresponds to a separation angle of slightly less than $1°$, whereas DMR's smallest angle was $7°$.

10. Details can be found on the Hubble Space Telescope (HST) Web site listed in the Bibliography. The actual age range given in the press release is 12–13 billion years, to which must be added an estimate of the time when the (ancient) white dwarfs were first formed, a time thought to be a little less than a billion years after the Big Bang. Hence the 13–14 billion year range quoted in the text.

11. In the language of note 9 to this chapter, the prediction was that as one went to larger values of ℓ, each succeeding peak height would be lower than its preceding neighbor's.

12. The parameters of Table 12 are from the collection of articles published by the WMAP collaboration as the Supplement to the *Astrophysical Journal* cited in the Bibliography.

13. Lahav and Liddle's reanalysis is reported in the Cosmological Parameters section of the 2004 Review of Particle Physics issue of *Physics Letters* cited in the Bibliography.

14. One way of estimating how much the presence of mass curves the space around it is to assume that the mass is concentrated at a point, as in the Schwarzschild radius discussed in Chapter 4. The point mass will warp space, producing a curvature C, the unit of which is inverse length squared $(1/[\text{length}]^2)$, for instance $1/\text{meter}^2$. Now let the body whose mass is warping space be a sphere of radius D. One estimate of the degree to which the body has distorted space is to compare the value of C at its surface with $1/D^2$, that is, by forming the dimensionless ratio $C/(1/D^2)$, a procedure discussed, for example, by Berry (1976). The ratios for a few representative bodies are approximately 7×10^{-9} (earth), 2×10^{-6} (sun), 10^{-3} (white dwarf), and 0.03 (neutron star). The first and second of these are perhaps surprisingly tiny, whereas the neutron star ratio is quite respectable. Since C is proportional to the mass M and inversely proportional to the radius D, the ratio is increased by simultaneously increasing the mass and decreasing the radius. For a black hole whose mass is spread over the event-horizon sphere, the ratio is 0.5, and as the sphere's radius becomes smaller and smaller than the Schwarzschild radius, the ratio will grow increasingly larger than 0.5. As this is my final descriptive comment about curvature, it is only fair to the reader to point out that there is a caveat not only to these comments but to the entire discussion concerning curvature, as well as the effect of mass warping space, Chapter 4. The caveat concerns the fact that in relativity (both the general theory and the earlier, more restrictive, special theory), space and time are inextricably linked, as stated in note 18 of Chapter 4. This linking leads to three-dimensional space being replaced by the four-dimensional entity denoted spacetime,

wherein time is considered to be a fourth dimension, analogous to the three spatial ones. Mass then distorts spacetime, not just space alone; correspondingly, it is spacetime that is curved. Hence, Figures 17 and 18 can more generally be thought of as representations of spacetime. For my purpose, it is simplest to have retained the spatial interpretation of these figures throughout. It is important to note that the curvature parameter k still enters the equations governing the time evolution of the scale factor R, and that the three types of geometry portrayed in Figure 23 apply to both space and to spacetime, though the visualization in the latter case is not quite as "straightforward." These are technical matters that only a reader with a great interest in them as well as a strong mathematics background will probably wish to pursue. Relevant material can be found in some of the technical literature cited in the Bibliography.

15. An enumeration of some of these tests can be found in Raine and Thomas (2001).

16. The argument outlined in this paragraph can be found, in more or less the same form, in the books by Harrison and Rich.

17. This value for the size of the visible Universe provides an *a posteriori* justification for the statement on page 133 that the Universe is so large that galaxies behave like structureless atoms. The structure is essentially invisible because the ratio of a typical galaxy's size to the diameter of the Universe is so small. Our Galaxy exemplifies this claim. Its diameter, from Chapter 5, is about 30 kpc. Dividing this number by the size of the visible Universe, one finds the ratio to be approximately 3.5×10^{-6}, not quite pointlike, but too small for any structural details to be discernable. This result may be compared with a similar ratio from atomic physics, that of nuclear sizes relative to those of atoms. The former is about 10^{-15} m, while the latter is about 10^{-10} m; the ratio is approximately 10^{-5}, a fraction small enough on the atomic scale that the neutron/proton structure of nuclei only plays a role in extremely precise measurements. See additional comments in Chapter 9.

Chapter 8: The Early Universe

1. Readers interested in events that occurred after the stages treated here, for example, the formation of galaxies, are recommended to the books by Harrison, Peacock, Raine and Thomas, Rees (2000), Rich, and Silk, references cited therein, and also the SDSS Web site.

2. The subscript "ph" in T_{ph} stands for "photon."
3. Energy units used in atomic physics are "electron volts," abbreviated eV; those employed in nuclear physics are millions of electron volts, written MeV; while in elementary particle physics one uses billions of electron volts, abbreviated GeV. One electron volt is the energy that a single electron or proton would gain if it experienced an electric potential of 1 V (one twelfth of the voltage between the terminals of the typical automobile battery—at least in the United States). The energy needed to ionize hydrogen is about 13.6 eV, whereas 2.2 MeV is needed to disintegrate a deuteron, the nucleus consisting of a neutron and a proton. The amount of energy contained in the mass of a proton is, from Einstein's $E = Mc^2$ formula, approximately 0.938 GeV, so that when a proton and an antiproton annihilate, about 1.88 GeV is released (as two photons). Since the sun has roughly 10^{57} protons, its mass energy is about 10^{57} GeV, whereas a kilogram of uranium (^{238}U) has a mass energy of about 4×10^{29} GeV, far less but enough to power a terrible weapon. Finally, the equivalence between energy and temperature is 1 eV $\cong 1.16 \times 10^4$ K and 1 K $\cong 0.86 \times 10^{-4}$ eV.
4. Discussions of the relationship between wavelength and the scale factor of an expanding universe can be found in the books by Harrison, Liddle, or Raine and Thomas.
5. The two proportionalities on page 181 can be re-expressed as the following equalities: $\lambda_{\mathrm{obs}} = C \times R_0$, and $\lambda_{\mathrm{emit}} = C \times R_{\mathrm{emit}}$, where the same constant of proportionality C occurs in each equation. Dividing the first of these two equations by the second gives $\lambda_{\mathrm{obs}}/\lambda_{\mathrm{emit}} = R_0/R_{\mathrm{emit}}$, and substituting this result into the expression for z, followed by a little rearranging, yields Equation (7) of page 181.
6. Each of the matter and radiation densities are averages taken over the volume of the (visible) Universe. Concentrating first on matter, this means that d_{M}, whether it refers to baryonic, dark, or both kinds of matter, contains the volume in the denominator: d_{M} is the ratio of mass to volume. Because volume has the dimension of distance cubed (it may be thought of as the product of height times length times width), and distance is proportional to R, then volume is proportional to R cubed (i.e., to R^3). Hence, the density of matter is inversely proportional to the cube of R: $d_{\mathrm{M}} \propto 1/R^3$. The density of matter rapidly increases as the Universe shrinks. Because density always contains volume in the denominator, it might be thought that radiation density d_{R} is also inversely proportional to the cube of R. However, radiation density is actually the ratio of radiation *energy* to volume, and the energy of radiation turns out to contain an inverse

factor of R. This is a result of Planck's formula, wherein the energy E_λ of a photon is inversely proportional to its wavelength λ: $E_\lambda \propto 1/\lambda$. As noted in the main text, wavelength in an expanding Universe is proportional to R, which means that E_λ is also inversely proportional to R: $E_\lambda \propto 1/R$. Thus, radiation density, being the ratio of radiation energy to volume, is proportional to the inverse fourth power of the scale factor: $d_R \propto 1/R^4$. Comparing with the behavior of d_M, it is seen that d_R increases even more rapidly than d_M as the mental time reversal leads to a continually shrinking Universe.

The constants that turn the preceding proportionalities into equalities are known for both d_R and d_M, so that the only variable is R. As one goes back in time, t.he decreasing value of R means that eventually $1/R^4$ will become larger than $1/R^3$. Inserting the constants of proportionality, one can then specify the particular value of R and, by implication, the exact time t_{eq} when the two densities were equal. That time has been calculated by the WMAP collaboration, and will be specified as an element of the timeline that is constructed later in the chapter.

7. This is another comment for the mathematically well prepared reader. The scale factor $R(t)$ has the following time dependencies: $R(t) \propto t^{1/2}$ for radiation domination, while $R(t) \propto t^{2/3}$ in the case of matter domination. These dependencies may be compared with the result for a universe dominated by the cosmological constant, given in note 2 of Chapter 7.

8. Since blackbody radiation contains photons of varying energies, recombination will occur over a finite period of time, not instantaneously. Correspondingly, there will be a spherical shell from which a whole series of last scatterings took place. Its thickness is estimated by Rich (2001) to be about one tenth the Hubble distance at recombination, or 20,000 pc.

9. The times of approximately 3 and 20 minutes aBB identifying the start and end of promordial nucleosynthesis are based on various numbers given by Weinberg (1977) and Rich (2001). Three minutes is Weinberg's approximation to 3 minutes 46 seconds, whereas Rich only estimates the start time as 3 minutes. He also uses a temperature estimate of approximately 0.03 Mev at which the reactions ceased (see note 3 to this chapter for a discussion of energy units and temperature equivalents). The time of 20 minutes (actually 20 minutes 25 seconds) aBB was obtained by expressing the 0.03 Mev temperature in Kelvin and assuming that both this temperature and the 7×10^8 K temperature corresponding to the onset of nucleosynthesis are blackbody. From this latter assumption, it follows that in

the radiation-dominated early Universe, $T_{ph} \propto 1/t^{1/2}$, a result arising from $T_{ph} \propto 1/R$, and $R \propto t^{1/2}$, as in note 5 to this chapter. The preceding blackbody assumption is not strictly correct, and therefore 20 minutes aBB is only an estimate.

Chapter 9: Conjectures

1. An equivalent use of this term occurs in mathematics, wherein a conjecture refers to a theorem that is unproved but whose validity is conjectured for reasons such as mathematical insight coupled with no known counterexamples and possibly profound consequences. An example is *Fermat's last theorem*, which states that if x, y, z, and n are integers, then the relation $x^n + y^n = z^n$ is true only when $n = 2$. The theorem was conjectured by the French mathematician Pierre de Fermat around 1637, intrigued amateurs and professionals for more than 350 years without either a valid proof or a negating counterexample having being found, and was finally proved in 1994 by the English mathematician Andrew Wiles. A history of this conjecture and its final proof that is accessible to readers of this book may be found in the monograph by Singh (1998).

2. That the wavelength of probes such as electromagnetic radiation should be comparable with the size of the object being probed may be understood as follows. First, recall that an object is seen when visible light is scattered from it, the scattered photons then entering the eye of the observer. From a macroscopic viewpoint, the scattered light wave has had its path disturbed by bouncing off the object, and it is the disturbed wave that gives rise to the observation. Next, consider water waves whose path could be disturbed by an obstruction in the water. Suppose that the obstruction is a small rock, and the wavelength of the water is much larger than the rock's size. In this case, the rock is too small to disturb the wave pattern, and the water will flow without being disturbed. If a disturbance in the pattern of the wave were the only means by which an observer could infer the presence of the rock, then he or she would not be able to tell if a small rock were in the path of the wave. It would be as if the rock were pointlike! The opposite would be true if the wavelength were much smaller than the rock's size: the pattern would change noticeably, and an observer could use it to infer the presence of the rock. While this analysis does not identify the minimum wavelength at which the rock will alter the wave pattern, it should be clear that it

will certainly happen when the two lengths are comparable, which is the statement made in the text.

3. The argument underlying this comment is that a photon's energy is inversely proportional to its wavelength ($E_\lambda \propto 1/\lambda$), and with the decrease in wavelength (so as to be able to investigate nuclear distances), the associated energy increases. The ratio of atomic to nuclear distances is roughly the ratio of nuclear to atomic energy scales.

4. In more technical language, microscopic systems whose particles interact *via* one or more of the four forces are said to be *invariant* under each of the symmetry operations. Such invariance means that applying a symmetry operation to the equations of an invariant theory that describe one system will lead to a physically allowed system but not necessarily the initial one. For example, your mirror image has left and right reversed, but the image is a possible person. However, it must be stressed that the results of identical experiments carried out before and after application of the symmetry operation will lead to *identical* results, as noted in the main text. A discussion of symmetries in elementary particle physics, at a level accessible to readers of this book, may be found in the book by Zee (1999).

5. The experiment detected two different decay sequences for the so-called neutral K meson, each of which behaved differently under the combination of charge conjugation and spatial reflection. The two decay modes can only occur if weak interaction theory fails to be invariant under the combination of the two symmetry operations. See Zee (1999).

6. The crucial element involved in the noninvariance of T is a mathematical theorem concerning the most fundamental theoretical framework in physics, known as *quantum field theory*. The theorem states that quantum field theory is invariant under the triple symmetry operation of C, P, and T, taken in any order. Let us call that triple symmetry operation CPT. Since CPT is always valid, and the combination CP is violated for the weak interaction, then T must be also, so that two "wrongs" will make CPT "right" again.

7. Readers who are interested in the various unified theories, quark–lepton interchange, GUTs, proton decay, the *supersymmetric* extensions of GUTs, etc., could consult Greene (1999), Guth (1997), Harrison (2000), or Zee (1999) for discussions of one or more of these topics at a level roughly that of this book. For those with a strong physics and mathematics background, the books by Raine and Thomas (2001), Rich (2001), or Peacock (1999) may be rewarding.

8. Although antimatter does not occur terrestrially and has not been found elsewhere in the Universe, it has been produced on the earth in a form other than particle–antiparticle pair creation. In 2002, anti-hydrogen—antimatter formed from an antiproton and a positron (the positively charged, antiparticle of the electron)—was produced at CERN, the European nuclear and high-energy research center.

9. 10^{-34} seconds aBB may seem like an unimaginably tiny number, but it is partially a matter of context. For example, the "size" of a proton is about 10^{-15} m, so that the time it takes a photon to cross that dis-tance is roughly 3×10^{-23} seconds. If it is taken as a microscopic time unit, then 10^{-34} seconds is "only" a trillionth of this unit. Small, but perhaps not unimaginably small.

10. Planck's constant is usually denoted h, not $\hbar$, the latter symbol being used instead for the "reduced Planck constant," viz., $\hbar = h/2\pi$, where π ($\cong 3.14159$) is the ratio of a circle's circumference to its diameter. $\hbar$ has been used in the text to represent Planck's constant in order to avoid any possible confusion with the empirical Hubble constant of Chapter 5. Expressed in the energy unit MeV, introduced in note 3 to Chapter 8, $\hbar$ has the approximate value of 6.58×10^{-22} MeV × seconds. An electron in one of the low-lying, excited quantum states of the hydrogen atom possesses a value of just $\hbar$ for an attribute known as *angular momentum*. By comparison, a ladybug walking radially at a speed of 0.01 m/sec at a distance of 0.1 m from the center of a turntable spinning at 10 revolutions per minute has an angular momentum of about $10^{28} \times \hbar$! Even for a macroscopic creature as tiny as a ladybug, $\hbar$ is a completely negligible quantity. It, and thus quantum theory, is important only in the context of microscopic phenomena.

11. For readers interested in how $\hbar$ enters formulas for length, energy, and time in quantum theory, here are a few instances. A length that typifies atomic sizes is the *Bohr radius*, denoted a_0. The formula for it is $a_0 = \hbar^2/M_e e^2$, where M_e is the electron's mass and e is the mag-nitude of the charge on an electron. a_0 is approximately equal to 0.5 $\times 10^{-10}$ m. The energy E_0 needed to ionize the hydrogen atom is $E_0 = e^2/2a_0$, whose value is approximately 13.6 eV. Finally, as an example of an atomic lifetime, consider the decay process in which an elec-tron in one of the lowest-lying excited states in the hydrogen atom makes a transition back to the ground state by emitting a photon. The lifetime, denoted here by τ, the lowercase Greek letter tau, is given by an expression too complicated to write out in detail, and I note only that τ is inversely proportional to the cube of Planck's con-stant: $\tau \propto 1/\hbar^3$ [$\tau \cong 1.6 \times 10^{-9}$ seconds].

12. Once again, for readers who wish to see the details, in the following I display the formulas for the Planck time, the Planck length, and the Planck energy in terms of $\hbar$, the speed of light c, and Newton's constant of gravity G. Using t, D, and E as the symbols for time, length, and energy, subscripted with Pl for Planck, the formulas are $t_{Pl} = (\hbar G/c^5)^{1/2}$, $D_{Pl} = (\hbar G/c^3)^{1/2}$, and $E_{Pl} = (\hbar c^5/G)^{1/2}$.

13. Brown dwarfs are instances of dark-matter candidates denoted MACHOS, an acronym for massive compact halo objects. The first term refers to the need for enough mass to make a significant contribution to Ω_0, the second to their need to be small and thus hard to detect, and the third one to the portion of galaxies where they are inferred to reside. The latter inference comes from both the galactic rotation curves and gravitational lensing, as discussed for example by Silk (2000). Other MACHO candidates are neutron stars and different types of black holes, but none seem likely to be a major component because of their presumed low density, as is also discussed by Silk (2000). At present, the preferred candidates are the exotic, weakly interacting, conjectured particles discussed in the main text. By the way, a third piece of evidence that strongly supports the need for dark matter comes from analysis of the CMB anisotropies. It shows that without dark matter, galaxies would not have had time to form, and therefore the observed large-scale structure in the Universe would not exist.

14. The uncertainty principle is a foundational element in quantum theory. It states that certain related pairs of microscopic quantities cannot each be simultaneously measured to arbitrary accuracy. Instead, as the accuracy increases in the measurement of one, the measured accuracy of the other will decrease, in contrast to macroscopic phenomena where any pair of measurements can be carried out to arbitrary accuracy. An example is the measurement of an electron's position and momentum (momentum is the product of the particle's mass and velocity). As applied to time and energy, the uncertainty principle is slightly different, stating, for instance, that in the context of pair creation, energy can be borrowed for a brief time. It is then subject to the requirement that the product of the borrowed energy times the time during which it is being borrowed must always be greater than a certain minimum amount. This subject is highly technical; curious, but very well prepared readers will find discussions in various textbooks and monographs, for example, that of the author, Levin (2002).

15. The concept of a nonempty vacuum, with pairs popping into and out of existence, may strike readers as science fiction. Far from it! It has

been shown, through elaborate and difficult calculations, that theory can yield agreement with certain experimental values (measured to extremely high accuracy) only by taking account of such "virtual" pair creation.

16. At the current time, string theory is another "hot" topic in both elementary particle physics and cosmology. Despite its formidable mathematics and rolled-up extra dimensions, it has been the subject of a variety of popular expositions, including Web sites, television programs, books, and articles in magazines like *Scientific American* or *Discover*. The mathematics involves quantities called "membranes," referred to by the practitioners as *branes*, and the theoretical ideas include exotic processes such as the collision of branes. These could possibly be the origin of the Big Bang. Much remains to be learned; not all theorists accept that string theory will turn out to be the ToE. Interested readers should check Web sites, etc. The book by Greene (1999) is an introduction aimed at the readership for whom *Calibrating the cosmos* was written; others exist and might be usefully read by those whose curiosity is whetted by the remarks herein. See the Bibliography for other sources of information.

17. These bi-gravity theories were proposed by T. Damour, I. Kogan, and A. Papazoglou in several technical articles published in 2002. This work is mentioned in the *Scientific American* article cited in the following chapter note.

18. That the acceleration could be due to the leakage of gravitons into the extra dimensions of string theory was proposed by Georgi Dvali and various collaborators in a series of technical articles. A description at a level suitable for interested readers of this book can be found in the February 2004 issue of *Scientific American*.

19. For a discussion of quantum gravity, in particular loop quantum gravity, see Smolin (2001).

20. The ekpyrotic theory was put forward by J. Khoury, B. Ovrut, P. Steinhardt, and N. Turok in several scientific articles, and later publicized, for example, in the February 2004 issue of *Discover* magazine. It is a conjecture that left some scientists skeptical because of its singular nature, and other scientists awaiting further developments.

21. One interpretation of a quantum phenomenon known as the *reduction* or *collapse of the wave packet* postulates the existence of other universes, although this *many worlds* interpretation is not so widely accepted. It is based on the probabilistic nature of quantum theory, which I briefly summarize in this long chapter note. I begin by noting that in the language of quantum theory, a microscopic system is "described" by an entity known as its *state vector*. What is meant

by "describe" and how the state vector does the describing is not germane, as will be seen.

To demonstrate how the wave packet collapses, let the quantum system be a photon that is to be detected on a screen after it is directed toward one of the two slits in Young's experiment, Chapter 3. The upper of the two slits will be identified as number 1, the lower as number 2. There will be a state vector describing passage of the photon through either one of the two slits, assuming the other is closed: These are state vectors I designate, respectively, as 1 and 2. Each of these two state vectors will give rise to a probability for finding the photon somewhere on the screen, still under the assumption that only one slit is open.

In Young's experiment, both slits were open, and in terms of the photon paradigm, it is not known which of the two slits any of the light beam photons passed through. In this situation, quantum theory decrees that the overall state vector describing detection of the photon when both slits are open is the *sum* of state vectors 1 and 2. This sum is the *wave packet* referred to above. Since it is the overall state vector that determines the probability of detection at any point on the screen, then each of ancillary state vectors 1 and 2 contribute to this probability. (Because light is made up of photons, it is the fact that the photon's overall state vector is a sum of the two ancillary ones that is responsible for the interference pattern observed when a beam of light, containing a huge number of photons, each described by the same overall state vector, is incident on the two-slit screen.)

Now I shall add a second detector to the mix. Its purpose is to identify which slit the photon passed through. This change in the experimental situation changes the state vector as well. In this changed situation, the overall state vector (the wave packet) *undergoes an instantaneous change* when the photon is identified as having passed through one of the two slits: the overall wave packet immediately turns into just the ancillary wave packet associated with passage through that particular slit. This instantaneous change is the reduction or collapse referred to above. The consequence is profound: instead of the probability being determined by both ancillary state vectors, it is now determined by only one of them (and in the experiment with a detector in place, the interference pattern first observed by Young disappears!).

This result has led quantum theorists to grapple with the following question: what happens to the other ancillary state vector when the change occurs? The answer provided by the many worlds interpretation is that at the moment detection forces the collapse, the Universe splits into separate and disjoint (totally disconnected) universes. For the two-slit/photon system, *our* Universe is the one in which the photon passed through the slit with the detector, whereas in the newly created universe the photon passed through the other slit!

Since all phenomena are quantal at some level, and many decisions are an analogue of choosing which slit the photon passed through (for example, does radiation cause a gene to mutate or not), the many worlds interpretation would lead to the generation of an enormous number of disjoint universes over time (whether humans were present or not). These are not, of course, cosmologically generated universes. Although this may seem more like science fiction than science, it is a legitimate scientific stance, though one not subject to verification, as far as is known.

It is a stance that raises amusing possibilities one might speculate about, somewhat along the lines of time travel in science fiction stories, though without their paradoxical side. For instance, a universe created by wave-packet collapse would, overall, be just a slightly changed copy of ours, although even tiny changes could have long-term consequences (does a butterfly flapping its wings in Asia cause catastrophic weather in Europe?). For example, in a newly minted, many worlds universe, particular people might either not be born or not have all of the same choices available to them, certain wars might not be fought, different political outcomes could occur, sporting events might turn out differently, etc. The speculative possibilities are enormous. One led to Michael Crichton's science fiction novel *Timeline*, later turned into a Hollywood film.

22. Rees (2000) points out that if the ratio were 0.006 or less, the deuteron would not be stable, and therefore no heavier nuclei could form, meaning stars could not burn nor could life as we know it exist. On the other hand, he notes that if the ratio were 0.008, two protons could overcome the Coulomb repulsion and bind together, in which case all protons would eventually be used up forming di-proton nuclei, leaving none available to form hydrogen atoms. Once again life as we know it would not have occurred.

23. How to interpret quantum theory and decide what it says about reality is a subject of vast content. Many books have been written about it, and curious readers might start by typing into their favorite Web browser phrases such as "quantum reality," "Schroedinger's cat," and "Bell's theorem" or inspect the physics section of a good bookstore.

Bibliography

Items marked with a single asterisk contain a mixture of technical and nontechnical material; the lay reader should usually have little difficulty understanding the latter portions. Items marked with a double asterisk are mostly if not entirely technical in nature and are listed for readers with a good to strong physics/mathematics background.

Publications

Abell, G. O. (1975): *Exploration of the Universe (Holt, Reinhart and Winston, New York)

Ashby, N. (2002): Relativity and the Global Positioning System, in *Physics Today*, **55**, 41

Berman, L. and Evans, J. C. (1977): *Exploring the Cosmos (Little, Brown, New York)

Bernstein, J. (1993): Cranks, Quarks and the Cosmos (Basic Books, New York)

Bernstein, J. and Feinberg, G., eds (1986): **Cosmological Constants (Columbia University Press, New York)

Berry, M. V. (1976): **Principles of Cosmology and Gravitation (Cambridge University Press, Cambridge, UK)

Born, M. (1957, with Blin-Stoyle, R. J): **Atomic Physics (Hafner, New York)

Brown, G. E. (2005): Hans Bethe and Astrophysical Theory, in *Physics Today* **58**, 62

Clayton, D. D. (1983): **Principles of Stellar Evolution and Nucleosynthesis (University of Chicago Press, Chicago)

Ferris, T. (1997): The Whole Shebang (Touchstone/Simon and Schuster, New York)

Feynman, R. P. (1985): QED (Princeton University Press, Princeton)

Gamow, G. (1970): My World Line (Viking Press, New York)

Greene, B. (1999): The Elegant Universe (Norton, New York)

Gregory, R. L. (1976): Eye and Brain (McGraw-Hill, New York)

Guth, A. H. (1998): The Inflationary Universe (Helix Books, Reading, MA)

Harrison, E. R. (2000): *Cosmology, 2nd ed. (Cambridge University Press, Cambridge, UK)

Koestler, A. (1959): *The Sleepwalkers* (MacMillan, New York)

Kolb, R. (1999): *Blind Watchers of the Sky* (Oxford University Press, Oxford)

Krauss, L. (2000): *Quintessence* (Basic Books, New York)

Kuhn, T. S. (1951): *The Copernican Revolution* (Harvard University Press, Cambridge, MA)

Lemonick, M. D. (2005): *Echo of the Big Bang* (Princeton University Press, Princeton)

Levin, F. S. (2002): **An Introduction to Quantum Theory* (Cambridge University Press, Cambridge, UK)

Liddle, A. (1999): **An Introduction to Modern Cosmology* (Wiley, Chichester)

Livio, M. (2000): *The Accelerating Universe* (Wiley, New York)

Miller, A. I. (2005): *Empire of the Stars* (Houghton Mifflin, Boston)

Pais, A. (1983): **Subtle is the Lord* (Oxford University Press, Oxford)

Park, D. (1997): *The Fire Within the Eye* (Princeton University Press, Princeton)

Peacock, J. A. (1999): **Cosmological Physics* (Cambridge University Press, Cambridge, UK)

Phillips, A. C. (1994): **The Physics of Stars* (Wiley, Chichester)

Physics Letters B **592** (2004), Issues 1–4, **Review of Particle Physics

Physical Review D **66** (2002), Part I, **Review of Particle Physics

Raine, D. J. and Thomas, E. G. (2001): **An Introduction to the Science of Cosmology* (Institute of Physics, Philadelphia)

Rees, M. (1997): *Before the Beginning* (Helix Books, Cambridge, MA)

———— (2000): *Just Six Numbers* (Basic Books, New York)

Rich, J. (2001): **Fundamentals of Cosmology* (Springer, Berlin)

Rigden, J. S. (2002): *Hydrogen* (Harvard University Press, Cambridge, MA)

Riordan, M. (1987): *Hunting of the Quark* (Simon and Schuster, New York)

Sanders, R. H. (2005): Observational Cosmology, in **The Physics of the Early Universe*, edited by E. Papantonopoulos (Springer, Berlin, Heidelberg)

Silk, J. (2000): *The Big Bang* (Freeman, New York)

Singh, S. (1998): *Fermat's Last Theorem* (Fourth Estate, London)

Smolin, Lee (2001): *Three Roads to Quantum Gravity* (Basic Books, New York)

Teresi, R. (2002): *Lost Discoveries* (Simon and Schuster, New York)

Thorne, K. S. (1994): *Black Holes and Time Warps* (Norton, New York)

Webb, S. (1999): *Measuring the Universe* (Springer Praxis, Chichester)

Weinberg, S. (1972): **Gravitation and Cosmology* (Wiley, New York)
———— (1977): *The First Three Minutes* (Basic Books, New York)
———— (1983): *The Discovery of Subatomic Particles* (Freeman, New York)
———— (1994): *Dreams of a Final Theory* (Vintage, New York)
Whittaker, E. T. (1951): ***A History of the Theories of Aether and Electricity*, Vol. I (Nelson, London; Dover, New York)
Zee, A. (1999): *Fearful Symmetry* (Princeton University Press, Princeton)

Web Sites

BBC Science News: news.bbc.co.uk/1/hi/sci/tech
BOOMERanG: cmb.phys.cwru.edu/boomerang
COBE: aether.lbl.gov/www/projects/cobe/
CBI: astro.caltech.edu/~tjp/CBI
DASI: astro.uchicago.edu/dasi/
Gene Smith's Astronomy Tutorial: cassfos02.ucsd.edu/public/astroed.html
High-Z Supernova Search: cfa-www.harvard.edu/cfa/oir/Research/supernova/highZ.htmlctio.noao.edu/~hzss/
Hubble Space Telescope (HST): hubblesite.org/newscenter/oposite.stsci.edu/pubinfo/pictures.html
HST Key Project: ipac.caltech.edu/H0kp/H0KeyProject.html
Inflation for Beginners: biols.susx.ac.uk/home/John_Gribbin/cosmo.htm
Loop Quantum Gravity: edge.org/3rd_culture/smolin03/smolin03_index.html
MAXIMA: cosmology.berkeley.edu/group/cmb
New Scientist: newscientist.com/quantum
Physics News Updates: Type this phrase into a search engine and then specify the year
Physics Web: physicsweb.org
SDSS: sdss.org
Solar Spectrum: jan.ncc.nau.edu/~gaud/bio326/class/individ/lesson1-2-.html
Supernova Cosmology Project: supernova.lbl.gov/public/
Supernova Dimming by Axions: t8web.lanl.gov/people/terning/axion.html
Superstring Theory: superstringtheory.com
Wayne Hu's Homepage: background.uchicago.edu/~whu/
WMAP: map.gsfc.nasa.gov/index.html

Glossary

aBB: Abbreviation for "after the Big Bang"

Absolute zero: The lowest possible temperature, and the zero of the **Kelvin** temperature scale

Alpha (α) particle: The **nucleus** of the normal helium atom, containing two **neutrons** and two **protons**

Angle of parallax: The measured angle in the method of **parallax**

Anisotropy: A deviation from perfect uniformity

Antimatter: Matter made up exclusively of **antiparticles**

Antiparticle: A particle whose **EM** properties are the opposite to that of the corresponding particle; examples are an **antineutrino**, an **antiproton**, or a **positron**

Apparent luminosity: The decreased brightness of a radiating object far away; its decreased energy radiated per unit time due to its distance away

Axion: A hypothetical particle, much lighter than a **neutralino**, postulated to be a component of **dark matter**

Baryon: Generic name for a **neutron**, a **proton**, or a **quark**

Beta decay: Typically the transformation of a **neutron** into a **proton**, an **electron**, and an **antineutrino**

Big Bang: The explosive event in the very early history of the Universe, when it was believed to have been tiny and very hot, which led to its current expansion and structure

Blackbody: A system of radiation and matter in which the latter emits as many **photons** as it absorbs

Blackbody curve or **spectrum:** The distribution over wavelengths or frequencies of the energy radiated by a **blackbody**

Blackbody radiation: The energy radiated by a **blackbody**

Black hole: One of the end stages of stellar evolution that can occur after a **supernova** explosion; the stellar remnant lives inside its **event horizon**

Brown dwarf: A failed star; one whose mass is too small to allow nuclear burning to occur

Carbon atom: The atom with six **electrons**, whose most commonly found **nucleus** has six **protons** and six **neutrons**

Cepheid variable: A particular type of pulsating star, whose **period** of pulsation is related to its **luminosity**

Charge conjugation: The formal process whereby a particle is transformed into its **antiparticle** and vice versa

Cosmic jerk: The transition from deceleration to acceleration of the Universe

Cosmic microwave background radiation (CMB): The relic, **blackbody** radiation from the early Universe, currently at a temperature of 2.725 degrees **Kelvin**

Cosmological constant: A constant that enters the equations of **general relativity**

Dark matter: Unidentified, nonluminous matter

Dark energy: Generic name for the unknown energy postulated to give rise to the acceleration of the Universe and possibly to the **cosmological constant**

Decoupling: The formation of neutral atoms (mainly **hydrogen**) in the early Universe, thereby allowing **photons** to flow freely; see **recombination**

Density: **Mass** per volume

Deuterium: A one-**electron** atom whose **nucleus** is a **deuteron**

Deuteron: The **nucleus** containing one **neutron** and one **proton**

Doppler shift: The change in **frequency** or **wavelength** of sound or radiation when either the emitter or the observer (or both) is in motion

Ekpyrotic model: A **string theory**–based framework that is postulated to explain the origin of the Universe

Electromagnetic force (or interaction): The force between charged particles due exclusively to their possessing charge

Electromagnetic (EM) radiation: The energy emitted (radiated) either by microscopic systems when they decay from a higher to a lower energy level or when a charged particle changes its velocity

Electromagnetic spectrum: The entire range of wavelengths or frequencies over which radiant energy occurs; see also **spectrum**

Electron: A negatively charged point particle, about 2000 times lighter than a **proton**

Electron degeneracy: A condition in which all the **electrons** in a **quantum system** are in their lowest **energy levels**

Electroweak theory: A theory that unites the quantum descriptions of **electromagnetic** interactions and the **weak interaction** governing decays of elementary particles

Energy level: One of the discrete energies in which a **quantum system** can be found

Energy spectrum: The collection of **energy levels** of a microscopic, **quantum system**

Event horizon: The surface of a sphere, created by the gravity of a point mass, from the interior of which almost nothing can escape

Fermion: Generic name for certain types of particles, of which **electrons, protons, neutrons,** and **quarks** are examples

Field: A disturbance in **spacetime** that is generated, for example, by mass (gravitational), or charge (electromagnetic) or an elementary particle (quantum)

Frequency: The repetition rate (number of recurrence times per second) of a **periodic** system

Galaxy: A very large collection of stars and gas held together by gravity

General relativity: Einstein's universe-generating theory of gravity

Gluon: A particle whose emission and absorption gives rise to the force between **quarks**

Grand unified theories (GUTs): Frameworks that unify **quantum chromodynamics** and **electroweak theory**

Gravitational force (or interaction): The force between bodies due exclusively to their possessing mass

Gravitational lensing: The bending of light due to the warping or distortion of space by a massive object such as a **quasar** or **galaxy**

Graviton: A hypothetical particle whose emission and absorption by masses gives rise to the **gravitational force**

Ground state: The lowest **energy level** of a **quantum system**

Hadron: A **baryon** or a **meson**

Hertzsprung–Russell diagram: A diagram on which stars are placed according to the values of their **luminosities** and temperatures (**spectral class**)

Higgs particle: A hypothetical particle in the **standard model** that causes the 16 others to have **mass**

Homogeneous (homogeneity): The property wherein no location can be distinguished from any other

Hydrogen atom: an atom containing a single **electron** and one **proton** in its **nucleus**

Kelvin: The unit in which **absolute** temperatures are measured

Ion: An atom or molecule in which **electrons** have been added or subtracted

Inflation: A theory postulating that the Universe increased enormously in size in a very short time very early in its history

Isotope: A **nucleus** differing from another only by the number of **neutrons** it contains

Isotropy (isotropic): The property wherein no direction can be distinguished from any other; also the condition of being perfectly uniform

Lepton: The generic name for any member of the family of **electrons** and **neutrinos**, plus their **antiparticles**

Light year: Approximately 9.45×10^{12} km, the distance light travels in 1 year

Loop quantum gravity: A quantum theory of gravity in which space and time each have minimum, rather than zero, values.

Luminosity: The energy per second radiated by a hot object, typically a star or **galaxy**

Main Sequence: The broad band of stars running from upper left to lower right on the **Hertzsprung–Russell diagram**

Mass: The quantity of matter in a body

Megaparsec: One million **parsecs**

Meson: A particle, made up of two **quarks**, that interacts *via* the **strong force** and that eventually decays into other particles

Neutralino: A hypothetical particle (much heavier than an **axion**) that is postulated to be a component of **dark matter**

Neutrino: An almost massless, neutral particle that is emitted when a **neutron** decays into it, a **proton** and an **electron**

Neutron: A neutral particle found in most **nuclei**; it is slightly heavier than a **proton**

Neutron star: A stellar end stage that can occur after a **supernova** explosion; the stellar remnant consists entirely of **neutrons**

Nova: A short-lived explosive event that occurs on the surface of a **white dwarf** star when it accretes matter from a **red giant** companion star

Nucleon: Generic name for a **neutron** or a **proton**

Nucleus (nuclei): The tiny, central core of an atom; it is composed of **protons** and (other than in the case of **hydrogen**) **neutrons**

Pair annihilation: An event in which a particle meets its **antiparticle** and is transformed from two masses into a pair of **photons**

Pair production: The creation of a particle and its **antiparticle** by converting the energy of radiation into mass

Parallax: A method for determining distance by observing an object from two vantage points separated by a known distance and then measuring the angle between the lines of sight to the object. One half of this angle is the **angle of parallax**

Parameter: A quantity that can take on different values

Parsec: 3.26 light years, which is the distance from the earth to an object whose **angle of parallax** is 1 second of arc

Period: The time it takes for an orbiting or other type of repeating/oscillating system to return to any point in its path

Photon: The massless, particle-like, discrete bundle or **quantum** of energy that constitutes **electromagnetic** radiation

Photosphere: The outer layer of a star, from which radiation is emitted into space

Plasma: A state of matter consisting of electrically charged particles and **photons**

Positron: The **antiparticle** to an **electron**

Power spectrum: A plot of quantities related to the temperature **anisotropy** *versus* the parameter ℓ, introduced in Chapter 7; see also endnote 9 of that chapter.

Primordial nucleosynthesis: Formation of very light **nuclei** in the early Universe

Proton: A positively charged particle found in all **nuclei**, slightly less massive than a **neutron**

Pulsar: A radiating, rotating **neutron star**

Quantum chromodynamics: The theory that describes the behavior of **quarks** and **gluons**

Quantum system: A microscopic body or system described by the theoretical framework known as quantum theory, which also describes some macroscopic systems such as **white dwarf** stars or **neutron stars**

Quark: One of a class of fundamental objects that are currently believed to be the only constituent of **neutrons** and **protons**; see Appendix C

Quasar: The supermassive **black hole** center of a highly luminous **galaxy**

Quintessence: A **quantum field** postulated to cause the acceleration of the Universe

Recombination: The formation of neutral atoms from **protons** and **electrons** (mainly **hydrogen**) in the early Universe, thereby allowing **photons** to flow freely; see also **decoupling**

Red giant: A large volume, low (surface) temperature stage into which a star of roughly the sun's **mass** evolves from its **Main Sequence** existence

Redshift: The increase in wavelength of radiation emitted when the source, the observer, or both are receding from each other

Redshift parameter: A quantity that measures the fractional change in **wavelength** of the radiation emitted by an object receding from the earth; it is denoted by the symbol z

Reionization: An event, initiated by the first **galaxies** and **quasars**, in which ultraviolet **photons** are absorbed by **hydrogen atoms**, breaking them up into their constituent **protons** and **electrons**

Scale factor: The single length characterizing a **homogeneous, isotropic** universe

Spacetime: The four dimensions made up of the three spatial ones plus time; the warping of spacetime by **mass** gives rise to **gravitational effects** in **general relativity**

Spatial reflection: The formal process by which lengths in any direction are turned into their mirror images

Spectral class: Classification scheme by which stars are identified *via* the **spectrum** of their emitted radiation

Spectrum: The collection of **energy levels** of, or **photons** emitted by, a body; see also **electromagnetic spectrum**

Standard model: The current framework that successfully describes the behavior of elementary particles. It comprises the **electroweak theory** and **quantum chromodynamics**

String theory: A mathematical framework in which elementary particles are tiny strings and which encompasses a quantum theory of gravity. Different versions of string theory are formulated in 10, 11 or larger numbers of dimensions

Strong force or **interaction:** The force between **quarks** and particles made up of them; it is mediated by the emission and absorption of **gluons**

Supernova: A violent stellar event in which an enormous amount of energy is radiated

Time reversal: The formal process whereby the positive flow of time is transformed into a reverse flow; running time backwards

Tunneling: A process whereby a **quantum system**, normally a particle, is able to pass into, through, and then out of a barrier

Wavelength: The smallest spatial distance between similar points on the oscillations of a periodic system

Weak force or **interaction:** The force responsible for decays of elementary particles; also the interaction between a **neutrino** and another particle

White dwarf: One of the three stellar end stages; it consists mainly of **carbon atoms**

List of Symbols

Symbol	Meaning
$\cong$	'Is approximately equal to'
$\propto$	'Is proportional to'
α	^{4}He nucleus, containing two neutrons and two protons
γ	Denotes either a photon or radiation
η	Ratio of the number densities of protons to photons
λ	Wavelength of a wave, a photon, or any radiation
ν; $\bar{\nu}$	A neutrino; an antineutrino
π	The ratio of a circle's circumference to its radius, approximately 3.14159
Λ	The cosmological constant
Ω	Relative density, relative to the critical density d_c
c	The speed of light, approximately 3×10^8 m/sec
cm	Abbreviation for centimeter
d	Density, or a deuteron, the nucleus with 1 proton and 1 neutron
d_c	The critical density, approximately 5 proton masses per cubic meter
e; e$^+$	An electron; a positron (antielectron)
f	Frequency
g	The acceleration of gravity at the surface of a massive body; relates mass and weight
$\hbar$	Planck's constant
k	The curvature parameter (= 1, 0, or –1)
kg	Abbreviation for kilogram
km	Abbreviation for kilometer
ℓ	Apparent luminosity; the CMB power spectrum parameter
ly	Abbreviation for light year
m	Abbreviation for meter
mi	Abbreviation for mile
n	A neutron
n	Number density
p	A proton
p	Pressure
pc	Abbreviation for parsec
sec	Abbreviation for second
t	Tritium, the nucleus with 2 neutrons and 1 proton
v	Velocity or speed
z	Redshift parameter
A	An angle
C	Charge conjugation operator
D	Distance
H	The Hubble parameter
H_0	The Hubble constant, equal to the value of H at the present time
Hz	Abbreviation for cycles/second

Symbol	Meaning
K	Denotes degrees expressed in the absolute temperature scale
L	Luminosity
M	Mass
Mpc	Abbreviation for megaparsec
P	Spatial reflection operator
R	The scale factor of a homomogeneous, isotropic universe
T	Temperature; time reversal operator
W	Abbreviation for watts
Y_p	Mass fraction (ratio) of the number of alpha particles to the total number of alphas plus protons

Author's Note and Acknowledgments

Three-plus years into retirement from Brown University, I developed an itch to teach again plus a desire to scratch that itch. However, after 31 years of teaching at Brown, that desire did not include assigning homework, making up exams, or any form of grading! The obvious means to achieving my goal was teaching in an adult-education setting. Not only would there be no grades, etc, but interest in the subject matter would be the sole motive for participation. My latent interest in cosmology and the universe having been reawakened by reading popular books on it in fall 2001, I proposed a course on this subject to the few institutions in my locale that offered an adult-education program. They responded positively to my proposal to teach a course for persons who had no science or mathematics background.

Nevertheless, while *my* background is physics, I am not a professional cosmologist: my research had been in the areas of nuclear reaction theory and collision theory. Creating the syllabus required an effort that expanded considerably when it came time to transform my lecture notes into this book. It has been an experience that has made the learning effort worthwhile many times over.

The syllabus was organized around a 16-hour teaching schedule (one 2-hour meeting per week for 8 weeks), one that would cover a set of topics intended to maintain the interest and enthusiasm of the adult students. Because the participants had little previous knowledge and were prepared to listen, the classes, originally planned as seminars, became straight lecture. However, in what was a reasonably successful attempt to promote class participation, I insisted that any one who felt uncertain about any aspect of the lecture should immediately stop me and ask for clarification. It may have helped in this regard that informality was the watchword in the classroom.

The course so far has been offered twice in the Circle of Scholars group at Salve Regina University in Newport, Rhode Island, and once at the South Coast Learning Network in New Bedford, Massachusetts. I am grateful to everyone at these places, most especially the class attendees, who helped make my plans a successful reality.

This book, and the lecture notes on which it is based, could not have been prepared without the articles and books in which I immersed myself and to which I am deeply indebted. The overall structure of *Calibrating the Cosmos* was influenced by that of Stephen Webb's *Measuring the Universe*, which has been a valuable source. Anyone familiar with his book will find traces of it in some of my early chapters (and sometimes more than traces). Important material in Chapter 4 came from Tony Phillips's *The Physics of Stars*, a book that I regret being unable to recommend to nontechnically trained readers.

My understanding of cosmology, and in particular of homogeneous, isotropic universes and of the early Universe, has benefited enormously from the books by Bernstein and Feinberg, Berry, Liddle, Harrison, Peacock, Raine and Thomas, Rich, and Webb; the 2002 and 2004 Reviews of Particle Properties; and numerous Web site materials. Rocky Kolb's *Blind Watchers of the Sky* was an enlightening source for historical and biographical information. I learned much from the popular books by Ferris, Livio, and Rees; their writing, like so much of that in the books already cited, has influenced my own presentation. The other books as well as the Web sites cited in the Bibliography have been helpful, as have a few older ones that remain uncited.

The book could not have reached its final form without the critical comments of family members, friends, and colleagues who read various drafts or portions of it. I am especially indebted and grateful in this regard to Eric Broudy, Jay Burns, Barbara Constance, Dorothea Doar, Antal Jevicki, Michael Levin, Jay O'Neil, Greg Tucker, and especially John Watson. It is a pleasure as well to acknowledge the splendid support provided by Harry Blom, Christopher Coughlin, Christopher Curioli, and Michael Koy of Springer Science+Business Media, LLC. None of the foregoing persons, of course, are responsible for any errors, infelicities, etc., that remain in the text. Finally, my lecture notes would almost certainly not have been turned into this book without the enormously motivating comments of my wife, Carol Levin, for whom the word "indebted" is totally inadequate, and to whom this book is lovingly dedicated. Thank you all.

Index

Printed in the United States of America.

RETURN TO: PHYSICS LIBRARY

351 LeConte Hall   510-642-3122